Y0-DNI-663

162
Advances in Polymer Science

Springer

Berlin
Heidelberg
New York
Hong Kong
London
Milan
Paris
Tokyo

Radiation Effects on Polymers for Biological Use

With contributions by
N. Anjum, Y. Chevolot, B. Gupta, D. Léonard,
H. J. Mathieu, L. A. Pruitt, L. Ruiz-Taylor, M. Scholz

 Springer

This series presents critical reviews of the present and future trends in polymer and biopolymer science including chemistry, physical chemistry, physics and materials science. It is addressed to all scientists at universities and in industry who wish to keep abreast of advances in the topics covered.

As a rule, contributions are specially commissioned. The editors and publishers will, however, always be pleased to receive suggestions and supplementary information. Papers are accepted for „Advances in Polymer Science" in English.

In references Advances in Polymer Science is abbreviated Adv Polym Sci and is cited as a journal.

Springer APS home page: http://link.springer.de/series/aps/ or
http://link.springer-ny.com/series/aps/
Springer-Verlag home page: http://www.springer.de

ISSN 0065-3195
ISBN 3-540-44020-8
Springer-Verlag Berlin Heidelberg New York

Library of Congress Catalog Card Number 61642

Springer-Verlag Berlin Heidelberg New York
a member of BertelsmannSpringer Science+Business Media GmbH
http://www.springer.de

© Springer-Verlag Berlin Heidelberg 2003
Printed in Germany

Typesetting: medio Technologies AG, Berlin
Cover: medio Technologies AG, Berlin
Printed on acid-free paper   02/3020kk - 5 4 3 2 1 0

ac

6-27 03
ac

Knl 7-7-03

Volume Editor

Prof. Dr. Henning Kausch
c/o IGC I, Lab. of Polyelectrolytes
and Biomacromolecules
EPFL-Ecublens
1015 Lausanne
Switzerland
E-mail: kausch.cully@bluewin.ch

Editorial Board

Preface

By polymers for biological use we understand biopolymers and living matter. Biomaterials are man-made or -modified materials which repair, reinforce or replace damaged functional parts of the (human) body. Hip joints, cardiovascular tubes or skin adhesives are just a few examples. Such materials are principally chosen for their mechanical performance (stiffness, strength, fatigue resistance). All mechanical and biological interactions between an implant and the body occur across the interface, which has to correspond as nearly as possible to its particular function. A natural surface is a complex (three-dimensional) structure, which has to fulfil many roles: recognition, adhesion (or rejection), transport or growth. We have to admit that at present biomaterials are far removed from such performance although new strategies in surface engineering have been adopted in which man tries to learn from nature.

Much of the progress in adapting polymer materials for use in a biological environment has been obtained through irradiation techniques. For this reason the most recent developments in 4 key areas are reviewed in this special volume. All surface engineering necessarily begins with an analysis of the topology and the elemental composition of a functional surface and of the degree of assimilation obtained by a particular modification. X-ray photoelectron spectroscopy (XPS) and time-of-flight secondary ion mass spectroscopy (ToF-SIMS) play a prominent role in such studies and these are detailed by H.J. Mathieu and his group from the Ecole Polytechnique Fédérale de Lausanne (EPFL). Generally, the first step towards procuring desired physico-chemical properties in a biomaterial substrate is a chemical modification of the surface. As pointed out by B. Gupta and N. Anjum from the Indian Institute of Technology (IIT), plasma- and radiation-induced grafting treatments are widely used since they have the particular advantage that they result in highly pure, sterile and versatile surfaces.

The sterilisation of implantable devices is a subject of great concern for the medical industry. Since ionising radiation is preferentially used for this purpose, attention must be paid to possible effects on the structural and mechanical properties of polymers (through chain scission or cross-linking). L. A. Pruitt from UC Berkeley has reviewed the specific behaviour of the different medical polymer classes to g- and high-energy electron irradiation and environmental effects. The biocidal efficiency relies on free radical formation and on the ability to reduce DNA replication in any bacterial spore present in a medical device.

The latter point, radiation effects on living cells and tissues, is the subject of the final contribution in this volume. M. Scholz from the Gesellschaft für Schwerionenforschung (GSI) summarises the (damaging) biological effects of ion beam irradiation and the considerable differences with respect to conventional photon radiation. These studies are of particular importance for radiation protection and radiotherapy. The advantages of a tumor treatment by carbon ion beams (effectiveness, concentrated energy release, possibility to use the presence of positron emitting 10C and 11C isotopes for positron emission tomography) are also presented in a comprehensive way.

I hope that the combination in a single special volume of the Advances in Polymer Science of these highly complementary contributions is particularly helpful to scientists working in this rapidly developing area. I would also like to thank all the authors for their exemplary co-operation.

Lausanne, December 2002   H. H. Kausch

Advances in Polymer Science
Available Electronically

For all customers with a standing order for Advances in Polymer Science we offer the electronic form via SpringerLink free of charge. Please contact your librarian who can receive a password for free access to the full articles. By registration at:

http://link.springer.de/series/aps/reg_form.htm

If you do not have a standing order you can nevertheless browse through the table of contents of the volumes and the abstracts of each article at:

http://link.springer.de/series/aps/
http://link.springer-ny.com/series/aps/

There you will find also information about the

– Editorial Board
– Aims and Scope
– Instructions for Authors
– Sample Contribution

Contents

Engineering and Characterization of Polymer Surfaces
for Biomedical Applications
H. J. Mathieu, Y. Chevolot, L. Ruiz-Taylor, D. Léonard 1

Plasma and Radiation-Induced Graft Modification of Polymers
for Biomedical Applications
B. Gupta, N. Anjum 35

The Effects of Radiation on the Structural and Mechanical
Properties of Medical Polymers
L. A. Pruitt 63

Effects of Ion Radiation on Cells and Tissues
M. Scholz 95

Engineering and Characterization of Polymer Surfaces for Biomedical Applications

Hans Jörg Mathieu[1] · Yann Chevolot[2] · Laurence Ruiz-Taylor[3] · Didier Léonard[4]

[1] Materials Institute, École Polytechnique Fédérale de Lausanne (EPFL), 1015 Lausanne EPFL, Switzerland. *E-mail: HansJoerg.Mathieu@EPFL.ch*
[2] Laboratoires Goëmar, UMR 1931 CNRS/Laboratoires Goëmar, Station biologique, 29660 Roscoff. France. *E-mail: chevolot@sb-roscoff.fr*
[3] Zyomyx Inc., 26101 Research Road, Hayward CA 94545, USA. *E-mail: LRuiz-Taylor@zyomyx.com*
[4] Analytical Technology, Microanalysis group, GE Plastics Europe, NL-4600 AC Bergen op Zoom, The Netherlands. *E-mail: V017704@gepex.ge.com*

The application of synthetic polymers in the growing field of materials for medical applications is illustrated by examples from recent work at the Materials Institute of the Swiss Federal Institute of Technology in Lausanne. The review highlights the need for functionalization and chemical control of material surfaces at a molecular/functional level. After a brief introduction into the surface chemical analysis tools, i.e., X-ray Photoelectron Spectroscopy (XPS) and Time-of-Flight Secondary Ion Mass Spectrometry (ToF-SIMS) combined with contact angle measurements, phosphorylcholine biomimicking polymers as well as immobilization of carbohydrates on polystyrene are presented.

Keywords: Polymers, Surface analysis, Functionalization, Immobilization, Glycoengineering

1	**Introduction**	3
2	**Methods for surface characterization**	3
2.1	Surface Chemical Analysis	4
2.1.1	X-ray Photoelectron Spectrometry (XPS)	4
2.1.2	Time-of-Flight Secondary Ion Mass Spectrometry (ToF-SIMS) . . .	8
2.2	Contact Angle Measurements	11
3	**Phosphorylcholine Functional Biomimicking Polymers**	13
4	**Surface Glycoengineering of Polystyrene**	23
5	**Concluding Remarks**.	31
	References	31

List of Abbreviations and Symbols

c_A	Atomic concentration of element A (at %)
Da	Dalton
e	Electron charge
E_b	Binding energy
E_{cin}	Kinetic energy
E_K	Core level binding energy
h	Planck's constant
v	Frequency of light
f_A	Isotopic abundance of element A
h	Height of sessile drop (angle of contact)
I_A	Intensity of XPS signal of element A
$I^{\pm}_A$	Intensity of positive or negative ions at mass A
I_p	Flux of primary ions
I_s	Flux of X-ray source
k	Instrumental variable (XPS)
K_{α}	X-ray transition
l	Diameter of droplet
L	Flight distance
m	Mass
M	Matrix
n	Number
S_A	Sensitivity factor of element A
S_i	Sensitivity factor of element i
T	Flight time
v	Speed
V_{acc}	Acceleration potential
z	Depth
Y_{tot}	Total sputter yield
$Y^{\pm}_{A(M)}$	Total positive or negative ion yield of element A in matrix M
z	Charge of molecule
ΔE	Energy resolution
Φ_A	Work function of analyzer (XPS)
$\gamma^{\pm}_{A(M)}$	Positive or negative ionization probability of element A in matrix M
γ_{lv}	Liquid-vapor interfacial tension
γ_{sv}	Solid-vapor interfacial tension
γ_{sl}	Solid-liquid interfacial tension
λ	Inelastic mean free path of photoelectrons
θ	Angle
θ_A	Advancing angle
Λ	Escape depth of photoelectrons
η_S	Analyzer constant (ToF-SIMS)

1
Introduction

Materials used in biomedical devices (so-called biomaterials) should fulfill two major requirements. First, they should possess the mechanical and physical properties that allow them to replace faulty functions of the body. Second, as they are interfacing biological systems they should be at least bio-inert (no undesired reactions) and, even better, trigger positive responses from these biological systems. In the latter case, responses are mainly governed by interfacial interactions, i.e., by the surface properties of the material such as surface energy, surface roughness, and surface chemical composition. Consequently, analytical methods are of primary importance to guide surface modifications of materials to lead to biospecific surface properties and also to understand the relationship between surface chemistry/roughness/energy and the biological response.

Although historically metals were the first biomaterials, polymers have gained a large application in this field and great efforts have been devoted to design polymeric materials with the right physical and interfacial properties [1]. This work intends to illustrate the application and importance of surface modifications and the need of surface analytical tools in the field of polymeric biomaterials [2, 3]. The examples given below stem from the work of several former PhD students and post-docs of our laboratory and illustrate how surface analysis and biological assays are used in a complementary manner to design successfully new soft biomaterials. The first example (Sect. 3) shows how one can synthesize a polymer whose surface properties reduce non-specific protein adsorption and consequently undesired biological reactions. The second example (Sect. 4) copes with the oriented immobilization of biologically active molecules (carbohydrates) at materials surfaces.

2
Methods for Surface Characterization [4]

In this first part, surface analytical methods such as X-ray Photoelectron Spectroscopy (XPS) and Time-of-Flight Secondary Ion Mass Spectrometry (ToF-SIMS) as well as contact angle measurements are briefly introduced. The first two techniques give chemical information on the first monolayers of a solid surface while the latter provides information related to the surface energy. In the following section, the basic principle of these analytical tools is discussed as well as the typical information they give and their limitations.

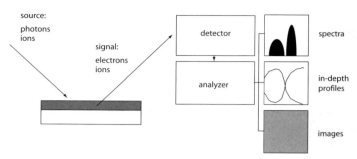

Fig. 1. Principle of XPS and ToF-SIMS surface analysis

2.1
Chemical Surface Analysis

Chemical analysis of surfaces and in particular of polymers is made possible by probing with photons (soft X-rays) or ions [5, 6]. The principle of chemical surface analysis is illustrated schematically in Fig. 1. A primary source is directed towards the surface of a solid sample and the spectrometer measures properties of emitted particles.

The emitted secondary particles (electrons or ions) carry information on the composition of the top-surface and underlying layers. In addition, the imaging capabilities make it possible in certain cases to identify heterogeneity in surface chemical composition. Also, the number of detected particles can be used for (semi-)quantification directly from the measured spectra, in-depth profiles, or images. More precisely XPS (probing with photons) and ToF-SIMS (probing with ions) allow one to access depths from 1 to 10 nm. However, due to their very high surface sensitivity, these methods are subject to contamination effects at the surface, requiring then a well-controlled preparation of the sample and ultra-high vacuum (UHV) conditions for analysis, i.e., pressures below 10^{-6} Pa. In the following a short description of the two analytical tools is presented.

2.1.1
X-ray Photoelectron Spectrometry (XPS)

XPS is a technique based upon photo-electronic effect. Under X-ray (photon) exposure, electrons are emitted with energy values characteristic of the elements present at the surface. Figure 2 illustrates the complete photo-ionization process including excitation and relaxation steps.

Let us assume that the primary X-ray with energy hν creates a photoelectron at the core energy level E_K (a). The Einstein equation gives the relation between exci-

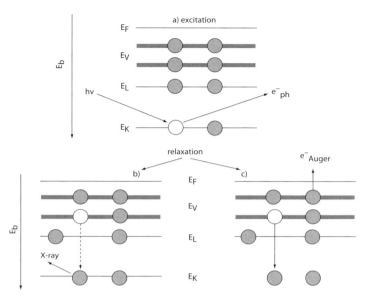

Fig. 2. Principle of the photo-electronic effect: the excitation and relaxation processes are shown indicating schematically the different binding states with the Fermi energy (E_F) as the reference level (=0). The valence band energy is E_V followed by a discrete level of E_L and the core level E_K; after [7]

tation energy ($h\nu$), kinetic energy E_{kin} of the emitted photoelectron and its binding energy E_b:

$$E_b = h\nu - E_{kin} - \Phi_A \qquad (1)$$

where Φ_A is the work function of the analyzer detector. An XPS spectrometer measures the kinetic energy E_{kin} of a core photoelectron. From Eq. (1), E_b is determined. The excitation process is exploited to identify solid elements from lithium (atomic number Z=3). The sensitivity limit in XPS is approximately 0.1% of a monolayer corresponding to 10^{15} particles/cm^2. An illustration is given in Fig. 3, which shows a survey spectrum of polystyrene (PS) after a radiofrequency oxygen plasma treatment. Typical reference energies for binding energy calibration are according to the ISO standard [8]:

Au 4f$_{7/2}$ 84.0 eV

C 1 s 285.0 eV

In Fig. 3, one observes the core level transitions of C1s and O1s as well as the Auger transition of oxygen, O_{KVV}. Indeed, the photoemission process is followed by a relaxation process either (b) or (c) as shown in Fig. 2. In the process (c) a third elec-

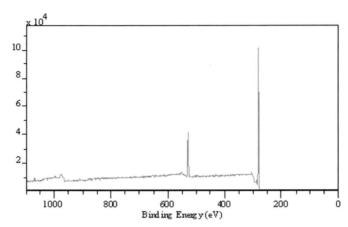

Fig. 3. XPS survey spectrum of a radiofrequency oxygen plasma treated polystyrene (PS) obtained with monochromatic Al primary radiation. The intensity I (cps) is shown as a function of binding energy (eV)

tron called the Auger electron is emitted after a transition of an electron from a lower level (for example the valence band level E_V) to the core level E_K. In the process (b), a secondary X-ray is created after filling the hole at the E_K level by a valence band electron. This photon is not measured in XPS experiments.

E_K can exhibit dependence upon the oxidation state of the element. Narrow scans around an element of interest allow one to determine quantitatively the various binding states of this element. In particular for polymers, carbon and oxygen binding states can be identified as illustrated in Fig. 4 for the C1s transition of PMMA.

Figure 4 shows the different functional groups and their respective relative areas are given in the caption. The theoretical relative intensities of the different functional groups are for carbon -C-C- plus -C-H (60%), -C-O-C- (20%), -C=O (20%) and oxygen -O-C- (50%) and -O=C- (50%), respectively.

XPS peak intensities (areas) I_A are a means of quantification. The relation between I_A and the atomic concentration c_A of an element or chemical component at a depth z is

$$I_A = kI_sS_A \int_0^\Lambda c_A(z)\exp\left(-\frac{z}{\lambda\cos\theta}\right)dz \qquad (2)$$

where k is an instrument variable, I_s the primary beam flux, S_A the elemental sensitivity, and λ the inelastic mean free path of the photoelectron trajectory multiplied by $\cos\theta$ where θ is the take-off angle of the emitted electron with respect to

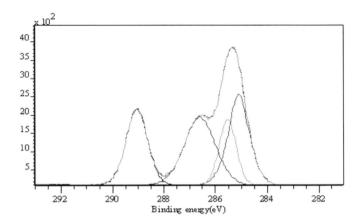

Fig. 4. XPS high resolution spectrum of C1s of Poly(methyl metacrylate) (PMMA). The four components correspond to C-H, C-C, C-O, and C=O functional groups of PMMA; after [9]

the surface normal. Due to the exponential term in Eq. (2) a reasonable upper limit for integration is

$$\Lambda = 3\lambda \cos\theta \tag{3}$$

Typical values of the escape depth Λ for polymers are between 5 and 10 nm indicating the shallow information depth of XPS [10]. Quantification is performed by application of the simple formula

$$c_A = \frac{I_A/S_A}{\sum\limits_{i=l}^{n} I_i/S_i} \tag{4}$$

applying elemental sensitivity factors from the literature (for example [5]) or those provided by the spectrometer manufacturers and summing over the number of elements taken into account. As for polymers accuracy of a few percent is typically obtained by use of Eq. (4). From Eq. (3) one can see that measurements at different take-off angles allow probing sample composition at different depths. Thus such angle resolved XPS (ARXPS) is an elegant way of obtaining a depth profile in a non-destructive manner.

When analyzing polymers, charging effects and possible degradation have to be taken into account [11–15]. Emitted photoelectrons carry a negative charge and may lead to a positive charge build-up. This effect can be compensated by supplying the sample surface with low energy electrons for charge neutralization. Degradation of polymer under X-ray is described in the literature. The reader is referred to [9–16] for complementary information.

Imaging is possible with a lateral resolution limit of a few microns for state-of-the-art spectrometers. All measurements are carried out under ultra high vacuum conditions (UHV). A fast entry lock is usually available for a transfer of a sample within minutes from atmospheric pressure to 10^{-6} Pa. Commonly two types of sources are used in XPS, either $MgK_{\alpha1.2}$ or Al $K_{\alpha1.2}$ radiation with an energy of 1253.6±0.7 or 1486.6±0.85 eV, respectively. Al $K_{\alpha1.2}$ radiation is often monochromatized by elimination of the $K_{\alpha2}$ ray to result in a better defined energy spread of the incoming X-rays, i.e., a smaller full width at half maximum (FWHM) of the K_α line allowing higher energy resolution ΔE down to 0.5 eV of the emitted photoelectrons. For more detailed information the reader is referred to the literature [17].

In conclusion, XPS and ARXPS are valuable tools for quantitative elemental analysis and identification of functional chemical groups within the first few nanometers of the surface at relatively high sensitivity.

2.1.2
Time-of-Flight Secondary Ion Mass Spectrometry (ToF-SIMS)

Static Secondary Ion Mass Spectrometry (S-SIMS) is a more sensitive (sensitivity of 10^9 atoms/cm^2) surface analysis technique than X-ray Photoelectron Spectroscopy (10^{12} atoms/cm^2) [6, 18]. Under primary bombardment with a focused ion beam the solid surface emits secondary particles. As illustrated in Fig. 5, the bombardment with primary ions such as Ga^+, Ar^+, or other molecular ion sources like SF_5^+ [19–21] or C_{60} [22], which should enhance molecular ion formation at high masses vs fragmentation, provokes the emission of neutral, positively, or negatively charged fragments and clusters.

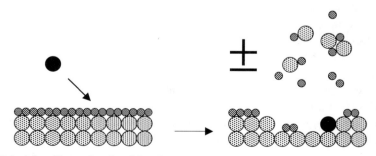

Fig. 5. Principle of Secondary Ion Mass Spectrometry

In SIMS, one distinguishes between the static and the dynamic modes. In static SIMS only a fraction of the first monolayer of the surface layer is perturbed. This depends on the flux of the primary ion beam, which is kept well below 10^{-12} particles/cm^2. This number corresponds to 1‰ of the number of particles of the first monolayer, which is approximately 10^{15} particles. Past this limit, degradation signatures can be detected. During further bombardment the top-surface will then be completely destroyed, leading to a depth profile type of information (dynamic mode).

The intensity $I_A{}^{\pm}$ for emitted positively or negatively charged secondary ions measured for a given target is described by the following equation:

$$I_A^{\pm} = I_p Y_{tot} Y_{A(M)}^{\pm} f_A c_A \eta_S \tag{5}$$

with

$$Y_M^{\pm} = Y_{tot} Y_{A(M)}^{\pm} \tag{6}$$

Here I_p is the primary ion current, Y_{tot} the total neutral sputter yield of species A in the matrix M, $\gamma_A{}^{\pm}(M)$ the ionization probability of species A in the matrix M, f_A the isotopic abundance of element A, c_A the atomic concentration, and η_S an instrumental constant. $I_A{}^{\pm}$ may vary over several orders of magnitude. Due to its high sensitivity, surface contamination may influence it strongly. Furthermore, $I_A{}^{\pm}$ is very dependent upon matrix effects that influence strongly the number of emitted ions compared to emitted neutrals. Their ratio is typically 10^{-4} or smaller for polymers and biomaterials [23]. The influence of the matrix on the ionization represents the severest limitation of this technique for quantitative analysis of both inorganic and organic materials.

In modern static SIMS instruments, the most efficient spectrometer is the time of flight spectrometer. It has typical current densities of the order of 1 nA/cm^2, which corresponds to approximately 10^{10} particles/cm^2/s [23, 24]. The emitted secondary ions are separated according to their mass in this spectrometer. Figure 6 shows schematically the principle of the time of flight measurement.

The relation between acceleration voltage, V_{acc}, and kinetic energy, E_{kin}, of the secondary ions is developed into

$$eV_{acc} = \frac{m}{2} v^2 \tag{7}$$

where e is the electronic charge, m the mass, and v the velocity of the accelerated ion. This leads to a simple relation between time of flight, t, and the square root of the ion mass:

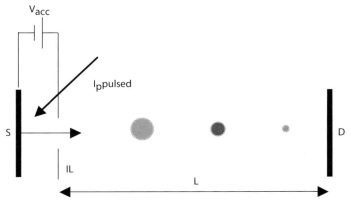

Fig. 6. Principle of a time-of-flight measurement: S sample, V_{acc} acceleration voltage, I_p pulsed ion beam, L flight length, IL immersion lens, D detector

$$t = \frac{L}{v} = \frac{L}{\sqrt{2eV_{acc}}}\sqrt{m} \tag{8}$$

where L is the flight distance being typically 1–2 m for commercial ToF-spectrometers. In ToF instruments the primary ion source is pulsed and the mass resolution will depend on the pulse width. The flight time for an ion with mass 100 Da is several microseconds and therefore picosecond pulses are required to obtain a relative mass resolution of m/Δm>5000 at m=28 Da. Liquid Metal Ion Guns (i.e., Ga^+ primary ions) allow one to focus the ion beam down to submicron beam diameter and then to perform SIMS imaging (mapping of elemental and molecular secondary ions). ToF-SIMS spectrometers are equipped with a fast entry lock allowing one to introduce a sample in the ultra-high vacuum range (UHV) within a few minutes.

Figure 7 shows schematically a spectrum acquired with a time-of-flight spectrometer. It shows a typical fragment of the maleimide molecule fragment $C_4H_2NO_2^-$ at 96.009 Da. Due to the mass resolution of m/Δm=6900 it can clearly be distinguished from the sulfate molecule SO_4^- at 95.952 Da. It illustrates that ion fragments can be measured with high mass resolution. This allows a unique determination of their chemical composition.

Examples of fingerprint spectra are shown in Sect. 3 (Figs. 11 and 12). Other detailed information is found elsewhere [17].

2.2
Contact Angle Measurements

Contact angle measurements are used to assess changes in the wetting characteristics of a surface to indicate changes in surface wettability. Information that one obtains largely depends on the interpretation of contact angle in terms of the Young equation [25, 26]:

$$\gamma_{lv} \cos\theta = \gamma_{sv} - \gamma_{sl} \tag{9}$$

where γ_{lv} is the liquid-vapor, γ_{sv} the solid-vapor, and γ_{sl} the solid-liquid interfacial tension, respectively, and θ the measured angle with respect to the surface, as illustrated schematically in Fig. 8.

The Young equation implies that the equilibrium contact angle is unique for a given solid-vapor-liquid system on a flat and rigid surface. In addition, it is generally accepted that the vapor spreading pressure can be neglected for $\theta > 10°$ [27]. However, in many practical cases the experimentally observed contact angle of a given system is not uniquely determined by the surface tensions of the solid and

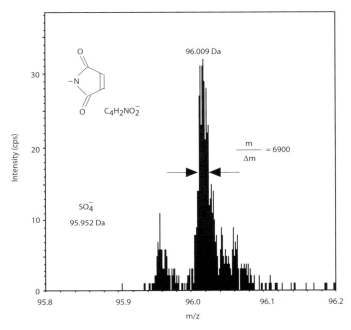

Fig. 7. Time-of-flight mass spectra of negative ions of $C_4H_2NO_2^-$ at 96.009 Da which is clearly separated from SO_4^- at 95.952 Da thanks to the mass resolution of $m/\Delta m = 6900$

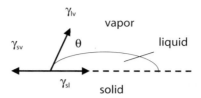

Fig. 8. Equilibrium sessile drop system; γ_{lv} is the liquid-vapor, γ_{sv} the solid-vapor, and γ_{sl} the solid-liquid interfacial tension, respectively and θ the measured angle with respect to the surface

liquid, but also by other parameters such as chemistry, inhomogeneity, roughness, surface deformation, surface reorganization, and chemical contamination [28, 29]. The current understanding of the drop size dependence of contact angles is discussed in a recent review by D. Li [29]. Surfaces are classified into high- and low-energy surfaces regarding their interfacial properties. For high-energy surfaces adsorption occurs easily, for low-energy surfaces not. Liquids and soft organic solids such as polymers exhibit surface energies below 100 mN/m, while for hard solids like metals it is around 500–5000 mN/m. A highly water- and oil repellent surface exhibits a very small critical surface energy as defined by Zisman [30]. Generally, surface energy decreases with increasing temperature and ambient pressure and rises with increasing salt concentration. A small contact angle results for wettable surfaces, if the interfacial energy is smaller that the surface energy of the pure material at the interface to air or vacuum.

Kovats et al. carried out substantial experimental work on contact angles and surface energy [31–35]. Different angle measuring methods were compared and surface tensions of 83 organic liquids determined. The importance of a reliable reference surface applying the Zisman concept is discussed in a theoretical contribution by Swain and Lipowsky [26] presenting a general form of the Young equation. Reviews of the critical surface energy determination are found elsewhere [36–38].

The influence of topography on wettability was discussed recently by Öner and McCarthy [39]. The hydrophobicity of a water repellent surface is discussed. In such a case the static contact angle is irrelevant, and the dynamic wettability has to be addressed by measuring the hysteresis, i.e., the difference between advancing and receding angle. The influence of roughness of ultrahydrophobic polymer surfaces (polypropylene and poly(tetrafluoroethylene) (PTFE)) exposed to a radiofrequency-plasma was discussed elsewhere [40] using XPS and Atomic Force Microscopy to determine size scale and topology of the roughness. Their most hydrophobic surfaces exhibited advancing (A) and receding (R) contact angles of θ_A and $\theta_R = 172°$ and 169°, respectively.

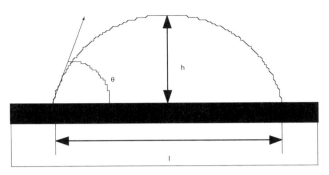

Fig. 9. Contact angle calculation; after [43]

Contact angle measurements of the advancing or receding angle can be performed under a microscope equipped with a CCD camera and a goniometer determining the slope of the droplet of a given volume of ultrapure water (= 10^{15} MΩcm). The wettability reported in Sect. 3 of this review was evaluated by a contact angle method using the sessile drop test based on the semi-empirical method proposed by the literature [41, 42]. The purity of the wetting agent was verified by measuring the liquid surface tension γ_{LV} using the Wilhelmy technique [28] (platinum plate, KSV sigma 70 Wilhelmy balance) and comparing the obtained value with the literature (γ_{LV}=78.8 mJ/m^2) [28, 43]. The height and the contact diameter l of a 1-μl drop of deionised water (grade nanopure milliQ, 17 MΩ) were determined after depositing the drop and taking a picture with a CCD camera [28]. The advancing angle θ was calculated using the following equation [43]:

$$\theta_A = 2\arctan\left(2\frac{h}{l}\right) \tag{10}$$

Height h and diameter l are illustrated in Fig. 9. They were determined after deposition of the drop and taking a CCD picture. The purity of the wetting agent was verified by measuring the liquid surface tension γ_{LV} using the Wilhelmy technique (platinum plate, KSV sigma 70 Wilhelmy balance) and comparing the obtained value with the literature (γ_{LV}=78.8 mJ/m2) [28, 43].

3
Phosphorylcholine Functional Biomimicking Polymers

This section illustrates results obtained in our laboratory in the context of the design and control of the surface properties of polymers used as biomaterials [43]. Indeed, interactions between biomaterials and tissues occur via a layer of proteins adsorbed at the surface of any implant [44]. Such protein adsorption is the imme-

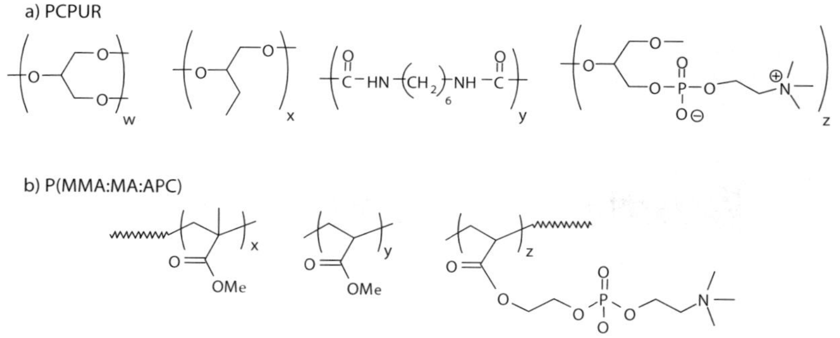

a) PCPUR

b) P(MMA:MA:APC)

Fig. 10a,b. Schematic structures of the PC copolymer synthesized: **a** PCPUR; **b** P(MMA:MA:APC); after [56, 58]

diate event occurring on its first contact with biological fluids and tissues [3]. One major restriction of polymers is their tendency to exhibit thrombogenic properties.

The response of biomolecules and cell membranes is determined by many factors, some of which are the chemical composition and conformation of the molecules, the surface energy, and topography of the top surface layers which are in contact with biological systems, i.e., body fluids and cells [45]. The work illustrated here consisted in designing new polymers with functional properties capable of promoting the attachment of specific cells. The first step consisted in a polymer system which surface inhibits non-specific cell attachment. This strategy is based on the incorporation of cell membrane constituents such as phosphorylcholine (PC) or phospholipid analogues into polymers [46–51].

Since polyurethanes have traditionally proved to be reasonably bio- and hemocompatible materials and have therefore been widely used for biomedical applications such as vascular prosthesis, artificial organs, blood contacting devices, peripheral nerve repair, or other prosthetic devices [52], our first PC copolymer systems were initially based on the polyurethane chemistry. We had earlier demonstrated that the presence of PC groups in poly(urethane) strongly reduces cell attachment at the surface even in protein enriched media [53] and work from Cooper and coworkers [54, 55] also showed that phosphorylcholine containing polyurethane limited neutrophil and bacterial adhesion. Two copolymers, PCPUR189 and PCPUR167, were synthesized with different concentration of PC moieties [56]. The final PC concentration in the bulk of the copolymers was 3.4 mol% and 4.3 mol% for the PCPUR189 and PCPUR167, respectively. Schematics of structure of the phosphorylcholine containing polyurethane (PCPUR) copolymers synthesized is given in Fig. 10.

The second copolymer system was chosen because of the higher synthetic flexibility offered by acrylate chemistry. Our choice was directed towards an acrylic terpolymer system, with methyl methacrylate (MMA) and methyl acrylate (MA) as principal components. The purpose was to enable the adaptation of the matrix mechanical properties by adjusting the glass transition temperature of the system [57]. 2-Acryloyloxyethyl phosphorylcholine (APC) was synthesized and added to MMA and MA through copolymerization to control the surface properties of the P(MMA:MA:APC) terpolymer with respect to cell attachment [58]. The final PC concentration in the bulk of the P(MMA:MA:APC) terpolymer was 1.7 mol%. A schematic representation of the structure of the terpolymer is given in Fig. 10.

A major part of these projects was also dedicated to understanding the bulk and solution structural organization of such polymers and here we would like to refer the reader to appropriate literature [56, 58]. While it was shown that the amphiphile nature of the terpolymer system, and in particular the phosphorylcholine (PC) groups, played an important role in the structure organization and molecular mobility of the copolymers, the results displayed in the present discussion focus on how the presence of these PC groups impacted on the surface properties of the copolymer synthesized. With this aim in mind, surface analytical techniques such as XPS and ToF-SIMS were used and complemented with contact angles and biological in-vitro assays to characterize the dynamics of the copolymers surface and assess the extent of the resistance to the non-specific attachment of cells on samples coated with the PC copolymers (for experimental details, refer to [56, 58]).

ToF-SIMS analysis of the 2-acryloyloxyethyl phosphorylcholine (APC) monomer led to the typical positive fragmentation pattern of the phosphorylcholine group in agreement with literature [59, 60]. Figure 11a illustrates the high relative intensity of some of these typical positive fragments such as $C_5H_{13}N+O_3P$ (166 Da) and $C_5H_{15}N+O_4P$ (184 Da). A more extensive list of the major positive ion-fragments detected is found elsewhere [58].

Figure 11b presents the same mass range of the positive SIMS spectrum obtained for the P(MMA:MA:APC) terpolymer. None of the typical fragments of the PC group can be identified despite the high surface sensitivity of the technique. In contrast, characteristic fragments from P(MMA:MA) matrix were observed in agreement with the literature [61], such as $C_8H_9O_3^+$ (153 Da) as well as $C_2H_3O_2^+$ (59 Da), $C_4H_5O^+$ (69 Da), and $C_5H_9O_2^+$ (101 Da) (not displayed in Fig. 11b). Negative ions spectra confirmed these results. Indeed, although the PO_3^- fragment could be detected, its intensity was too low to consider this as evidence of PC presence.

In contrast, investigations of PCPUR polymers using ToF-SIMS show that both polymers exhibit similar positive and negative mode mass spectra. In this case, major secondary ions characteristic of the PC polar head group were observed (PO_3^- (79 Da), $C_5H_{12}N^+$ (86 Da), $C_5H_{13}NO_2P^+$ (150 Da), $C_5H_{13}NO_3P^+$ (166 Da), and $C_5H_{15}NO_4P^+$ (184 Da) as shown in Fig. 12a,b.

Furthermore, the characteristic fragment of GPC containing coatings, $C_8H_{19}NO_4P^+$ (224D), was also detected (data not shown). The comparison of the

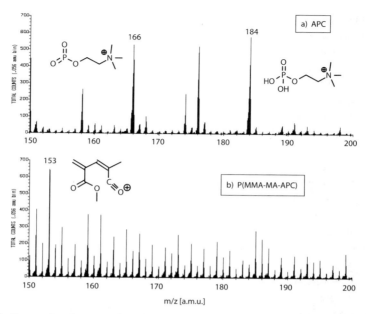

Fig. 11a,b. Comparison between the positive ToF-SIMS mass spectra of the APC: **a** monomer; **b** terpolymer in the 150–200 Da mass range; after [58]

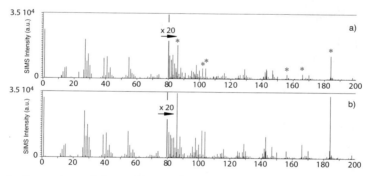

Fig. 12a,b. Comparison of the positive mode ToF-SIMS spectra of: **a** PC-PUR189; **b** PC-PUR167 in the mass range 0–200 Da, after [56]; * denotes the position of the characteristic PC fragments

positive mode ToF-SIMS spectra shows, however, that relative normalized intensities of the characteristic PC fragments are higher by approximately 50% for PCPUR167 compared to PCPUR189.

Studies regarding the conformation of lecithins have allowed the determination of the size of the phosphorylcholine polar group. This dimension derived from X-

Table 1. AR-XPS surface elemental concentrations of PCPUR and P(MMA:MA:APC) copolymers. Comparison between two emission angles of 20 and 80 ° with respect to the sample surface, corresponding to an information depth of 3–4 nm and 8–11 nm, respectively; compiled after [56, 58]

Emission angle (information depth)		C [at%]	O [at%]	N [at%]	P [at%]
PCPUR189	20 ° (3–4 nm)	67.1±0.5	21.4±0.8	11.5±0.5	0.3±0.09
	80 ° (8–11 nm)	66.1±0.0	21.6±0.3	11.7±0.2	0.6±0.06
PCPUR167	20 ° (3–4 nm)	66.7±0.6	21.6±0.7	11.4±0.3	0.6±0.06
	80 ° (8–11 nm)	65.9±0.8	21.8±0.7	11.7±0.3	0.8±0.06
P(MMA:MA:APC)	20 ° (3–4 nm)	70.0±0.6	29.8±0.6	0.0	0.2±0.02
	80 ° (8–11 nm)	70.0±0.0	29.4±0.0	0.3±0.06	0.3±0.03

ray diffraction studies is 11 Å for both the monohydrated or fully hydrated lecithins [62] and is in agreement with the 12 Å obtained with space-filling models for a fully extended PC group parallel to the fatty acid chains and with the zwitterion in the gauche conformation about the O-C-C-N bonds [63]. Hence, the typical fragments of the phosphorylcholine groups should be detectable if they were present at the uppermost surface of the P(MMA:MA:APC) terpolymer. Possible matrix effects have already been reported in organic systems [64, 65]. Although they may affect the emission probabilities of typical PC secondary ions, it is very unlikely that they should inhibit completely the emission of all the PC fragments. These SIMS observations therefore suggest that under the UHV conditions required for the analysis, the extreme surface of the P(MMA:MA:APC) terpolymer is depleted in PC groups. This burying effect can directly be understood by the desire of the system to reorganize so as to minimize surface energy in UHV environment.

XPS analyses were performed on all copolymers. The atomic concentrations of carbon (C_{1s}), oxygen (O_{1s}), nitrogen (N_{1s}), and phosphorus (P_{2p}) were compared between the two depths of 3–4 nm and 8–11 nm and are reported in Table 1.

Contrasting with ToF-SIMS analysis, nitrogen and phosphorus were detected on the P(MMA:MA:APC), although in the case of nitrogen we were close to the detection limit. As it can also be seen from the high resolution elemental scans reported in Fig. 13a,b, the nitrogen and phosphorus atomic concentrations measured are very low; for an emission angle of 20° no nitrogen is even detected whereas the phosphorus concentration is about 0.2 at%. This can be understood by considering the difference of electron inelastic mean free paths (imfp) between both elements. Indeed, calculations based on the estimation of the imfp made by Tanuma et al. [66] for PMMA as well as on the relation derived by Seah and Dench [67] for the imfp of organic compounds allow the estimation of the escape depth of the XPS information. Both approaches give similar results. In the case of nitrogen, the information depth is approximately 3 nm and 9 nm whereas for phosphorus it

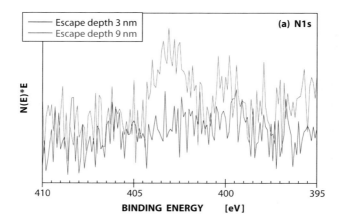

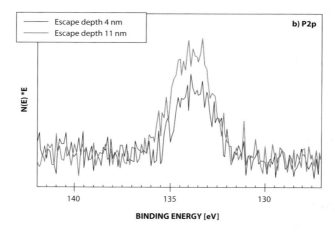

Fig. 13a,b. High-resolution elemental scans as a function of depth of: **a** the nitrogen N_{1s}; **b** the phosphorous P_{2p} of the P(MMA-MA-APC) copolymer; after [58]

ranges from 4 nm to 11 nm for the 20° and 80° emission angles, respectively. As the information depth is augmented, an increase in the nitrogen (from 0 to 0.3 at%) and phosphorus (from 0.2 to 0.3 at%) atomic concentrations is noticeable.

Similar comments can be made concerning the PCPUR copolymers, for which elemental concentrations are comparable for both copolymers at both emission angles, except in the case of phosphorus. Indeed, although phosphorus concentrations detected are very low, they are significantly different (outside of the experimental deviation) and show that the concentration is higher in PCPUR167 (Table 1) in agreement with the ToF-SIMS results described earlier. Moreover, it can be seen that the phosphorus P_{2p} photoelectron intensity (and consequently the phosphorus atomic concentration) also increases with the depth analyzed for these PCPUR copolymers.

This suggests that under UHV conditions, for both copolymer systems (acrylates or polyurethanes), there is an increasing concentration gradient of the phosphorylcholine groups from the surface to the bulk of the polymer. The copolymer system tends to lower its interfacial free energy by burying the PC group underneath the extreme surface.

In-vitro assays were performed in serum free (SFM) and serum containing media (SCM) to evaluate the cell attachment properties on both PC-containing and PC-free copolymers together with SiO_2 as reference [56, 58]. Results obtained after 4 h of incubation (Figs. 14 and 15) show that the cell attachment and differentiation levels on the PC containing copolymers were strongly reduced in both media compared to PC free system and the SiO_2 reference. Hence, a small concentration of PC groups (~1.7 mol% for the P(MMA:MA:APC) and ~3.4 mol% for PCPUR189) is already efficient in reducing the non-specific attachment of cells by roughly 70% for both media. Cell attachment is even further reduced as the PC content of the copolymer increases (~4.3 mol% for PCPUR167), since a 90% decrease is obtained for PCPUR167 in both media when compared to SiO_2.

Extension of the culture time up to, respectively, three days for the P(MMA:MA:APC) copolymer and four days for the PCPUR copolymers, showed no evolution in the cell attachment properties on the PC containing surfaces, as opposed to the PC-free surfaces (SiO_2 reference as well as P(MMA:MA) polymer). Indeed, after 72 h in SCM, a large population of cells were attaching and differentiating on P(MMA:MA) as can be seen from Fig. 16a. In contrast, on the

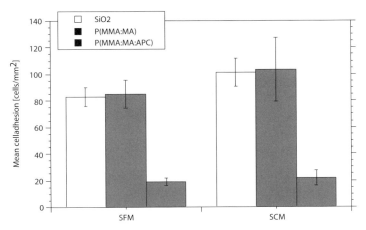

Fig. 14. Comparison of the cell attachment level after 4 h incubation, on the APC free and the APC containing copolymers as a function of time in serum free (SFM) and serum containing (SCM) media; after [58]

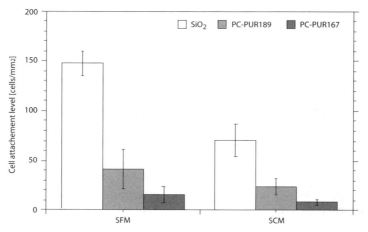

Fig. 15. Comparison of the cell attachment level on the SiO_2 reference, PC-PUR189 and PC-PUR167 after 4 h in SFM and SCM (*p<0.01 PCPUR190/PCPUR167 Student t-test, n=15 and **p<0.001, PCPUR189/PCPUR167 Student t-test, n=15) ; after [56]

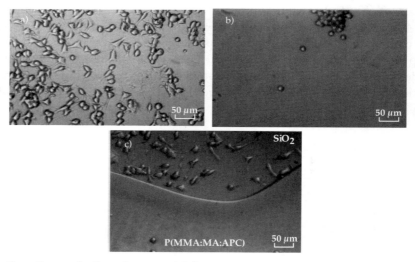

Fig. 16a–c. Extent of cell attachment and differentiation on: **a** P(MMA:MA); **b** P(MMA:MA:APC); **c** SiO_2, after 72 h in serum-containing media [58]

P(MMA:MA:APC) surface, cells rather aggregate to form clusters than adhere to the terpolymer surface (Fig. 16b), or preferentially grow on the uncovered silicon part of the sample (Fig. 16c). This last observation gives additional information about the absence of any toxic leakages from the acrylic terpolymer film. The cell attachment level was comparable in both SFM and SCM suggesting that the pres-

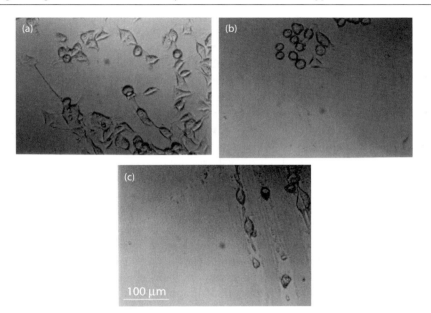

Fig. 17. Extent of cell attachment after 96 h in serum containing media, on: **a** SiO$_2$ reference substrate; **b** PCPUR189 in SCM; **c** PCPUR167 in SCM. Note the extensive reduction in cell attachment on PCPUR polymers compared to the reference substrate; after [56]

ence of binding proteins did not significantly influence the cell attachment level on the acrylic terpolymer surface.

Figure 17 illustrates the extent of cell attachment on the PCPUR copolymers after 96 h in the protein rich media (SCM). Although few cells are attaching on PCPUR189 (Fig. 17b) and have adopted this typical triangular morphology indicative of good attachment, neurite extension on this substrate is restricted compared to the SiO$_2$ reference (Fig. 17a). Furthermore, on PCPUR167 samples, almost no adhering cells remain, except on a small area of the sample where the PCPUR167 coating was accidentally removed. Cells are then growing on the bare SiO$_2$ substrate along the scratched portion of the coating (Fig. 17c).

Non-specific adsorption of various proteins to these copolymers was also studied [56] and followed essentially the same trend as the one observed for the non-specific attachment of cells, i.e., the non-specific protein adsorption was lower on the copolymer with the highest PC content. Hence, this inhibition of cell attachment on PCPUR polymer surfaces (and non-specific protein adsorption) can be directly related to the presence of PC groups. This implies that under aqueous conditions the PC groups were now present at the extreme surface for the copolymers to display such non-permissive properties.

Static water wettability measurements performed on PCPUR189 and PCPUR167, showed that the mean contact angle of a drop of water was 20° smaller on PCPUR167 (60±3.5°) compared to PCPUR189 (81±2.5°). This suggests an increased hydrophilicity of PCPUR167 (enhanced water-polymer interaction), which is in agreement with the higher phosphorylcholine groups concentration. In addition, in contrast to PCPUR189, the drop was observed to spread and flatten at the surface of the PCPUR167 polymer. Such a phenomenon can be attributed to both swelling and chain reorientation [68, 69], which appears easier in the case of PCPUR167 in agreement with the water weight gain and DSC observations [56].

Results of XPS and ToF-SIMS characterizations correlated to in-vitro assays and contact angles obtained for both phosphorylcholine containing polymer systems (PC-PURs and P(MMA:MA:APC)) have shown that the amphiphile nature of those copolymers and in particular the presence of the PC moieties generates a highly mobile outermost surface capable of quickly reorganizing as a function of the polarity of the environment it is subjected to. Under UHV conditions, both systems tend to reorganize so as to minimize their interfacial free energy by burying the PC groups, which have no affinity for the UHV environment and hence lowering the PC groups surface concentration. Many similar examples of surface reorganization of polymer systems upon the polarity of the environment have been reported [69–73]. In the case of the acrylate system, the extreme surface analyzed with ToF-SIMS was even depleted in PC groups. A difference in matrix effects is possible between the polyurethane and acrylate systems and can make a quantitative comparison of the PC content difficult. However, it is still not expected that different ions should be observed for both systems. Indeed, these fragments follow the same fragmentation process, regardless of the matrix they are emitted from; they could be observed for GPC, PC-PURs, APC, as well as for phospholipids [56, 59, 60]. ToF-SIMS results show that the relative intensity of the PC fragments decreases, as the PC bulk concentration decreases from 4.3 mol% in PC-PUR167 to 3.4 mol% in PC-PUR189, until fragments are no longer detectable at the outermost surface when the PC bulk concentration is 1.7 mol% in the acrylate system. On the other hand, AR-XPS measurements have shown that the PC groups concentration augments as the depth of analysis increases.

In the case of the acrylate system, the PC depletion of the outermost surface of the terpolymer suggests that there can be a PC threshold concentration below which, under UHV conditions, PC groups can be completely buried/hidden from the surface. However, this PC threshold concentration may not only be depending on the PC bulk concentration of the polymer. Indeed, the difference in polarity between the matrix and the PC groups is greater in the P(MMA:MA:APC) acrylate system than in polyurethane polymers. The driving force for surface reorganization is, therefore, greater in the P(MMA:MA:APC) terpolymer since the polar PC groups have no affinity for the UHV environment and are masked by hydrophobic polymer chains. On the contrary, in response to a polar environment, the terpolymer surface rearranges to present the phosphorylcholine functionalities at the top surface, hence minimizing interfacial energy and exhibit strong non-permissive properties with respect to cell attachment. The amphiphile nature of the terpolymer constitutes by itself the main driving force for surface reorganization. It generates a highly mobile polymer surface, with quick self-reorganization properties upon the environment. Besides, cohesion energy of the different constituents of the acrylate terpolymer (and especially of the hydrocarbon backbone) is always lower than the energy barrier to rotation of a single C-C bond (12.6 kJ/mol) [74] meaning that chains are free to rotate, hence promoting surface reorganization.

In the case of PC-PURs polymers, surface reorganization certainly also occurs although, even if as suggested by the wettability behavior of the polymers, it was not as obvious as for the acrylate system. A PC group concentration difference close to 50% could already be detected by ToF-SIMS at the outermost surface between PC-PUR189 and PC-PUR167. Furthermore, the PC concentration, as observed using AR-XPS, increased with the analyzed depth. Nevertheless, the driving force for surface reorganization is certainly smaller, due to lower polarity differences between PC groups and the polyurethane matrix. The large amount of water retained in this polymer structure [56] also leads to an apparently more polar matrix. Besides, the ability of the urethane and urea segments of the matrix to form hydrogen bonds between N-H and C=O groups increases the cohesion energy of the system, thus restricting the molecular mobility of the polymer chains [74].

In-vitro experiments correlated to bulk and surface characterization results have demonstrated that an increase of PC groups concentration such as in the polyurethane system leads to a reduction of the cell attachment level in both serum containing and serum free media. The presence of phosphorylcholine groups, as demonstrated on the acrylate system, is clearly responsible for the non-permissive properties exhibited by the PC containing surfaces with respect to cell attachment. Moreover, a low PC concentration (≈ 1.7 mol%) is already sufficient to reduce strongly cell adhesion at the surface of the terpolymer. Bulk organization (due to higher polarity differences) between the phosphorylcholine groups and a "flexible" polymer matrix seems to be the determining factor in providing a very compliant surface that can easily adapt to its environment. Hence, the high compliance of the surface and the dynamic flip-flop behavior of the PC groups are certainly playing a crucial role in the prevention of the non-specific adsorption of proteins and cells on such PC containing copolymers.

4
Surface Glycoengineering of Polystyrene

A derivatization can be performed to induce a selective attachment of target cells either on new polymers designed to control non-specific cell attachment (Sect. 3) or on any other substrate intended for biomaterial applications. Photoimmobilization of peptidic sequences was performed on newly synthesized PCPUR con-

taining polymers described in the section Phosphorylcholine Functional Biomimicking Polymers above [56]. This section of this review illustrates the work done in
our laboratory on the surface glyco-engineering [75], i.e., the specific covalent attachment of carbohydrates at the surface of substrates via photo-addressable newly synthesized molecules. It is wished to emphasize how surface analytical tools
and biological assays give complementary information on (1) the immobilization
of intact molecules and (2) the availability of active molecules for recognition.

Biological systems make considerable use of surface glycosylation. Indeed glycosylated molecules are involved in, among others, molecule trafficking, cell recognition (including species discrimination), blood group typing, and blood coagulation cascade [76–79].

Carbohydrate immobilization is reported in the literature for galactose [80],
melibiose [81], mannose [81, 82], lactose [82] starch [83], lactosaminide [84], and
heparin [85–89]. This has been achieved mainly by thin film adsorption (weak interactions) or by covalent binding. In our work, light was used as a means for oriented covalent immobilization of carbohydrates. Orientation and availability are
key parameters to allow the interaction of the cell receptor with the immobilized
molecules. Photoimmobilization provides a versatile tool with respect to the substrate (organic and inorganic) and allows one to create easily microdomains of bi-
orecognition with addressable printing, mask-assisted lithography techniques.
The photoreagents most often used for photoimmobilization of biomolecules are
arylazides, trifluoromethyl-aryl diazirines, and benzophenones [90]. These rea-
gents generate very reactive intermediates upon light activation. Their interaction
with the support material leads to the formation of covalent bonds [90]. Thus, the
synthesis of molecules containing carbohydrate and photoactivatable domains
makes it possible to achieve covalent binding of biologically active carbohydrates.
Utilizing diazirine as a photoactivatable function, a reactive carbene is generated
by thermochemical or light activation (350 nm). A covalent bond is generated
with the surface provided that close contact between the surface and the carbene
is obtained. Reaction of aryl diazirine with various substrates has previously been
described [91–94]. The carbene may insert into C-H, C-C, C=C, N-H, O-H, S-H
bonds [90]. Chevolot et al. [95] reported the synthesis and the immobilization on
a CVD deposited diamond surface of an aryl diazirine containing a galactose pho-
toreagent {4-(3-D-galactopyranosylsulfanyl-2,5-dioxopyrrolidin-1-yl)-N-(3-(3-
trifluoromethyl-3H-diazirin-3-yl)phenyl}butyramide (MAD-Gal) (Scheme 1)
[95]. Based on XPS and ToF-SIMS data it was concluded that grafting of the photoactivatable reagent MAD-Gal is possible using glycosylated aryldiazirines. Using
a masking technique [96] a specific pattern of immobilized carbohydrate was laid
down on diamond and verified with ToF-SIMS. This is illustrated in Fig. 18.

Léonard et al. [94, 96, 97] demonstrated that the molecule was immobilized as
a whole without degradation due to the immobilization process. They also demonstrated that insertion on diamond was indeed carbene mediated. This was evi-

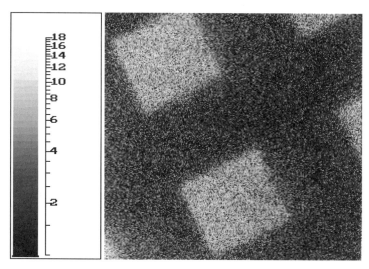

Fig. 18. Negative-mode ToF-SIMS of F- after mask-assisted patterning on thin diamond films – size of the image 115∞115 μm; after [75]

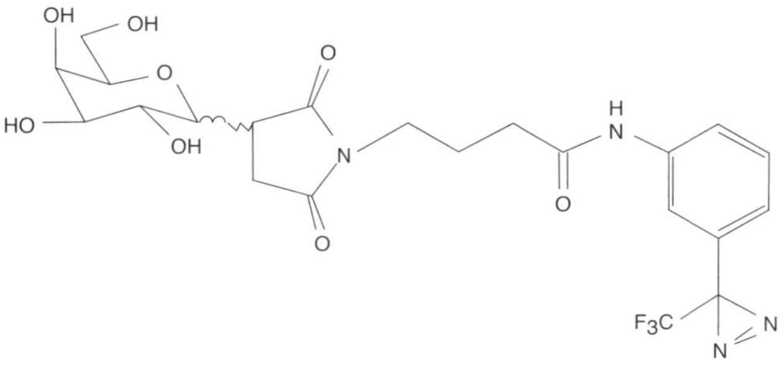

Scheme 1. Structure of MAD–Gal

denced in the case of oxygenated sites at the surface of CVD deposited diamond but other functional sites could also be involved.

In the following discussion, it is intended to compare the immobilization of two photoactivatable carbohydrates on polystyrene (PS) and to illustrate how their biological activity was tested. The same galactose aryl diazirine as the one described for diamond modification (see Scheme 1) was used and compared to the also newly synthesized lactose aryl diazirine displayed in Scheme 2.

The immobilization on polystyrene (the material most often used in biological assays) was controlled with surface analysis spectroscopic methods XPS and ToF-

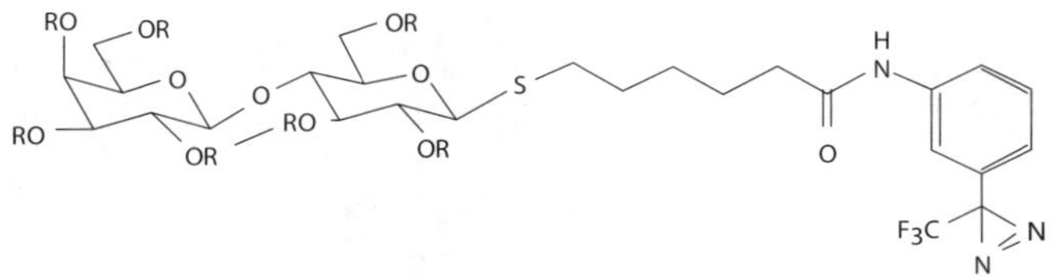

Scheme 2. Chemical structure of lactose aryl diazirine (R=H)

SIMS. The biological activity of the polystyrene-modified surfaces was probed with the lectin Allo A, primary rat hepatocytes, and α-2,6-sialyltransferase.

Allo A lectin was reported to have a specific affinity for galactose residues especially as part of lactose (3.1 mmol/l inhibits hemoglutinin activity) and O-nitrophenyl β-D-galactopyranoside (inhibitory at 12.5 mmol/l) [98]. Hepatocytes express on their surface the asialoglycoprotein receptor which is responsible for the clearance of abnormal galactose-terminated serum glycoproteins [99, 100]. Most serum glycoproteins carry terminal sialic acid residues and a penultimate galactose residue. When desialylated, the exposed galactose residues of the glycoprotein can interact with the asialoglycoprotein receptor, initiating removal of the glycoprotein from the circulation by endocytosis. Subsequently, the incorporated protein is hydrolyzed in lysosomes [100–103]. The transfer of sialic acid by α-2,6-sialyltransferase [104] to galactose attached to a solid support was studied to develop the approach to solid phase semi-synthesis [105].

Immobilization on PS was carried out as described by Léonard et al. [94, 96]. The samples were washed four times in water (HPLC grade). In the following experimental discussion, samples are referred to as A, B, and C. Sample A is the pristine polystyrene. Sample B is polystyrene with photoactivatable carbohydrates after glycosilation and washing. This sample treatment does not include light exposure. Physisorbed molecules should be removed by washing. Sample C is the polystyrene surface with molecules deposited on it, then photoactivated with light of 350 nm wavelength and finally washed leaving an expectedly modified PS surface.

When comparing samples A, B and C, simple signatures of the molecules such as fluorine, CF_3 among others, should allow one to test easily the efficiency of their immobilization on PS with XPS and ToF-SIMS. Table 2 illustrates the case of lactose aryl diazirine. The residual level of molecules observed on sample B by ToF-SIMS was below the XPS instrument detection limit. It confirms that the washing procedure was able to remove physisorbed molecules. For sample C, after washing, covalently attached molecules were expected to remain bound to the surface, in contrast with sample B [94, 96]. XPS fluorine atomic percentage, as well as CF_3^- and F^- normalized intensities (Table 2), illustrate in the same way the successful immobilization of the molecule at the surface of polystyrene. These results are

Table 2. XPS and ToF-SIMS analysis of lactose aryl diazirine grafted PS. XPS atomic percentages are displayed for samples **A**, **B** and **C**. Sample **A** corresponds to as received polystyrene. Sample **B** corresponds to polystyrene on which a methanolic solution of lactose aryl diazirine was deposited, and no light activation was performed before the final washing. This surface should be similar to sample **A**. Sample **C** corresponds to polystyrene on which a methanolic solution of lactose aryl diazirine was deposited and light activation was performed before the final washing. The molecule should be present at the surface of the material. Three areas were analyzed per sample, after [105]

XPS percentages (at %)	Sample A	Sample B	Sample C
N	bdl	0.34±0.20	0.51±0.49
F	bdl	bdl	1.18±0.4
S	bdl	bdl	0.18±0.16
C	97.79±0.36	94.53±1.27	88.77±0.54
O	2.27±0.47	5.13±1.08	9.35±0.49
ToF-SIMS			
Corrected total intensity (counts) $\infty 10^4$	16.3±4.5	49.6±5.4	43.7±10.7
F$^-$ normalized intensity ‰	1.10	2.72±0.42	63.17±21.78
CF$_3^-$ normalized intensity ‰	0.13	0.20±0.08	0.65±0.07

similar to those of the immobilization of MAD-Gal on diamond [96] and on PS [106].

Lactose and galactose residue bioavailability at the surface of polystyrene was probed with the β-galactose specific lectin Allo A through a biotinylated Allo A (Fig. 19). The amount of surface lectin was measured with ^{35}S streptavidin. 100% radioactivity relates to the total amount of radioactivity of the deposited radiolabeled streptavidin before washing. This represents an arbitrary standard but is used to indicate the relative levels of radioactivity remaining after the various treatments of the modified surface. The results are illustrated in Figs. 19 and 20 for lactose and galactose modification experiments of the polystyrene surface, respectively. For the two molecules, the percentages of radioactivity for samples **A** were similar (1.2 and 2.5%). For samples **B** (lactose aryl diazirine), the signal remained very low. For samples **C** (lactose aryl diazirine), the radioactivity reached the value of approximately 3%, which is seven times lower than for the corresponding sample **C** in the case of MAD-Gal experiments as shown in Fig. 20.

Furthermore, inhibition with asialofetuin of lectin binding on immobilized galactose (MAD-Gal) was observed [105]. It confirmed that the interaction was specific and not due to a variation in physisorption, which in turn is related to the surface energy.

In addition, XPS demonstrated that the surface concentrations of the two molecules (MAD-gal, lactose aryl diazirine) were almost similar (see Table 2). Figure 21 indicates quantitative XPS and normalized ToF-SIMS data of MAD-Gal

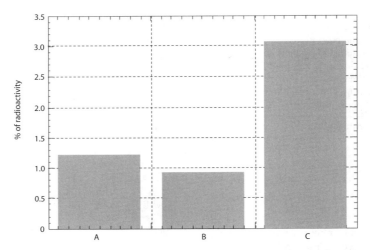

Fig. 19. Binding of biotinylated Allo A lectin to aryl lactose derivatized PS surfaces. Derivatized surfaces were incubated with Allo A lectin and then extensively washed to remove physisorbed lectins. [^{35}S] streptavidin was incubated and the surfaces were again rinsed to remove excess streptavidin. Finally, radioactivity was measured by scintillation counting. The radioactivity is higher on sample C; after [105]

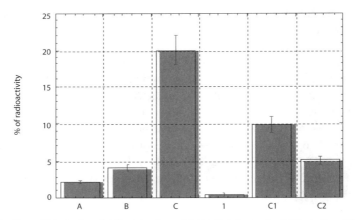

Fig. 20. Binding of biotinylated Allo A lectin to MAD-Gal derivatized PS surface. The derivatized PS surfaces were incubated with Allo A lectin and then extensively washed to remove physisorbed lectins. [^{35}S] streptavidin was added and the surfaces were again washed to remove excess streptavidin. Finally, radioactivity was measured by scintillation counting. The radioactivity is higher on sample C and increases with the concentration of the MAD-Gal solution (C, C1 and C2 corresponding to 0.25, 0.025 and 0.0025 mmol/l, respectively) used for immobilization. This illustrates the higher binding of streptavidin to the surface **C**. I corresponds to incubation of MAD-Gal grafted PS with asialofetuin and Allo A lectin. 100% corresponds to the total added radioactive streptavidin; after [105]

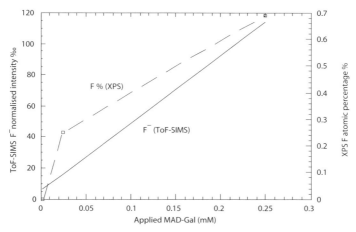

Fig. 21. F⁻ normalized intensity values (absolute intensity of F⁻/(total intensity – H⁻ intensity)) and fluorine atomic percentages are displayed as a function of the MAD-Gal concentration used for immobilization. The intensity of surface characteristic signatures increases with increasing density of MAD-Gal; after [105]

immobilized on PS as a function of the MAD-Gal concentration used for immobilization.

Therefore, the difference in lectin binding can be attributed to differences in affinity of the lectin for the different terminal galactose residues at the surface. According to Yamashita [98] the affinity of the lectin was observed to decrease for various substrates such as phenyl β-D-galactopyranoside, O-nitrophenyl β-D-galactopyranoside, lactose, and lactulose. It is therefore possible that in this case and as for phenyl β-D-galactopyranoside and O-nitrophenyl β-D-galactopyranoside, the succinimidyl group enhanced the affinity of the lectin towards the galactose of MAD-Gal relative to the galactose residue of the immobilized lactose.

On the cellular level, functionality of the galactose and lactose residues was tested with primary rat hepatocytes. The MAD-Gal coated surface altered hepatocellular function. MTT (3-(4, 5-dimethylthiazol-2-yl)-2, 5-diphenyltetrazolium bromide) and Neutral Red (NR) uptake (Fig. 22) increased for hepatocytes cultured on MAD-Gal coated polystyrene surfaces and the increase correlated with the density of galactose molecules on the surface. MTT formation and NR uptake increased up to a concentration of 2.5 mmol/l galactose and showed no saturation effect. CMF stands for crude membrane fraction which is taken as a reference.

NR uptake is related to lysosomal activity while MTT is related to the bioreductive uptake, when lectins also specific of galactose residues (RCA) were added most probably because lectins bind to the terminal galactose residues of MAD-Gal. In the case of lactose modified polystyrene, no increase in the NR uptake was observed (data not shown). On the other hand, when surfaces are then incubated

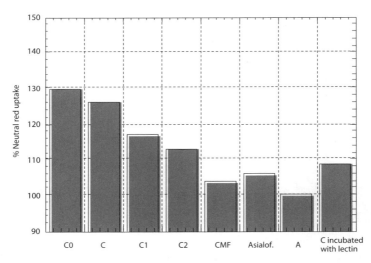

Fig. 22. Neutral Red (NR) uptake is related to the lysosomal activity. The NR uptake decreased with the decreasing of surface galactose residues (C0, C, C1, C2 corresponding to 2.5, 0.25, 0.025, and 0.0025 mmol/l). The NR uptake was similar to that of surface A (plain polystyrene) when the surface was incubated with a lectin (RCA) that protects the galactose residues. Hepatocytes possess higher NR uptake on asialofetuin coated surfaces than on surface A, but lower than on MAD-Gal grafted PS. 100% corresponds to surface A; after [105]

with ^{14}C CMP-NeuAc and α-2, 6-sialyltransferase, radioactivity was only detected on lactose modified PS (not illustrated).

It was also demonstrated that the radioactivity on the surface increased with incubation time. When incubated without the enzyme, radioactivity was similar to the background level. These results illustrate that the enzyme was able to recognize lactose on PS surface as a substrate for the transfer of sialic acid and to catalyze this transfer [105].

Improvement of the biological interactions will require more complex structures (cluster, spacer design, oligosaccharides). For example, as in the case of asialoglycoprotein receptor, a tri-anternary structure of the galactose residues increases the affinity of the receptor for the blood circulating glycoprotein. It was also demonstrated that the distance between the residues of the tri-anternary structure influences the affinity [105]. In a similar manner surface glycoengineering will require such an architectural design.

From the presented data it is concluded that diazirine-containing photoreagents containing mono- and disaccharides can be synthesized and immobilized on polystyrene. Biological activity of the modified polystyrene was demonstrated with Allo A lectin, primary rat hepatocytes and α-2, 6-sialyltransferase (not shown here) [105]. It was also demonstrated that the biological activity was not only dependent on the terminal residue but probably also on the spacer structure. Nevertheless, the combination of chemistry, surface analytical tools, and biology is found to be a powerful feedback system for the design of bioactive glycoengineered surfaces.

5
Concluding Remarks

This review illustrates the efficiency of the surface modification of polymers with bio-specific properties allowing key-lock interactions with cells. An advanced chemistry is needed to design new polymers exhibiting control of non-specific cell attachment allowing further surface bio-specific surface interactions. New components can be designed permitting bio-active surface modifications like photoaddressable carbohydrate compounds. Furthermore, surface analytical tools like XPS and ToF-SIMS are important for the preparation and control of the intended surface properties. Their spectra and images give insight into the top-surface composition with a high sensitivity. They can even be directly correlated to biological results such as the increase of MAD-Gal surface density correlated with the increase in the observed biological response. The complementarity of both techniques is also illustrated by these examples. The difference in probing depth allows highlighting compositional differences at the surface as well as gaining information on functional group orientation. This review illustrates that surface analysis techniques are powerful analytical means for the control of bio-active polymer engineering. The high sensitivity (ToF-SIMS) and the possibility to quantify data (XPS) identify these techniques as a major tool for further developments of biomaterials.

Acknowledgements. The authors would like to thank N. Xanthopoulos, G. Coullerez, and X. Gao for help with some of the experiments and discussions. Financial support is acknowledged from the Swiss Priority Program of Materials and the Common research program in Biomedical Engineering 1999–2002 between the University Hospitals of Geneva and Lausanne (HUG and CHUV), the Universities of Geneva and Lausanne, and the Swiss Federal Institute of Technology Lausanne (EPFL) as well as from the Bundesamt für Berufsbildung und Technologie (BBT/KTI), Bern, contract no. CTI-5170.1 MTS.

References

1. Ratner BD, Chilkoti A, Lopez GP (1990) Plasma deposition and treatment for biomaterial applications. In: d'Agostino R (ed) Plasma deposition, treatment and etching of polymers. Academic Press, San Diego, pp 463–516
2. Williams DF (1992) Biofunctionality and biocompatibility. In: Williams DF (ed) Medical and dental materials. VCH, Weinheim, pp 1–27

3. Hubbell JA, Langer R (1995) Chem Eng News 73:42–54
4. Mathieu HJ (2001) Surf Interface Anal 32:3–9
5. Briggs D, Seah MP (1990) (ed) Practical surface analysis, vol I – Auger and X-ray photo-electron spectroscopy, 2nd edn, vol 1. Wiley, Chichester
6. Briggs D, Seah MP (1992) (ed) Practical surface analysis, vol II – ion and neutral spectroscopy, vol 2. Wiley, Chichester
7. Mathieu HJ (2001) Elemental analysis by AES, XPS and SIMS. In: Alfassi ZB (ed) Non-destructive elemental analysis. Blackwell Science, Oxford, pp 201–231
8. ISO (2001) Surface chemical analysis – X-ray photoelectron spectrometers – calibration of energy scales. In: International Standard 15472, 2001
9. Beamson G, Briggs D (1992) High resolution XPS of organic polymers. The Scienta ESCA300 database, Wiley, Chichester
10. Powell CJ, Jablonski A (1999) J Phys Chem Ref Data 28:19–62
11. Frydman E, Cohen H, Maoz R, Sagiv J (1997) Langmuir 13:5089–5106
12. Buchwalter LP, Czornyj G (1990) J Vac Sci Tech A 8:781–784
13. Chaney R, Barth G (1987) Fresenius J Anal Chem 326:143
14. Clark DT, Brennan WJ (1986) J Electron Spectrosc 41:399–410
15. Storp S (1985) Spectrochim Acta B 40:745–756
16. Coullerez G, Chevolot Y, Léonard D, Xanthopoulos N, Mathieu HJ (1999) J Surf Anal 5:235–239
17. Vickerman JC (ed) (1997) Surface analysis – the principal techniques. Wiley, Chichester
18. Briggs D (1992) Static SIMS-surface analysis of organic materials. In: Briggs D, Seah MP (eds) Practical surface analysis, vol 2 – ion and neutral spectroscopy. Wiley, Chichester, pp 367–423
19. Benninghoven A, Stapel D, Brox O, Burkhardt H, Crone C, Thiemann M, Arlinghaus HA (1999) Static SIMS with molecular primary ions. In: SIMS XII. Elsevier, Brussels
20. Stapel D, Thiemann M, Hagenhoff B, Benninghoven A (1999) Secondary ion emission from LB-layers under molecular primary ion bombardment. In: SIMS XII. Elsevier, Brussels
21. Hagenhoff B, Cobben PL, Bendel C, Niehuis E, Benninghoven A (1997) Polymers under SF5 bombardment – a systematic investigation. In: SIMS XI. Wiley, Orlando, Florida
22. Justes DR, Harris RD, Stipdonk MJV, Schweikert AE (1997) A comparison of Cs and C60 primary projectiles for the characterization of GaAs and Si surfaces. In: SIMS XI. Wiley, Orlando, Florida
23. Vickerman JC, Briggs D (eds) (2001) ToF-SIMS: surface analysis by mass spectrometry. IM Publications and Surface Spectra, Chichester
24. Vickerman JC, Swift AW (1997) Secondary ion mass spectrometry. In: Vickerman JC (ed) Surface analysis – the principal techniques. Wiley, Chichester, pp 135–214
25. Young T (1805) Philos Trans R Soc London 96:65
26. Swain PS, Lipowsky R (1998) Langmuir 14:6772–6780
27. Garbassi F, Morra M, Occhiello E (eds) (1994) Polymer surfaces from physics to technology. Wiley, Chichester
28. Good RJ (1993) Contact angle, wetting, and adhesion: a critical review. In: Mittal KL (ed) Contact angle, wettability and adhesion. VSP-Wiley, Weinheim, pp 3–36
29. Li D (1996) Colloid Surf A 116:1–23
30. Zisman WA (1964) Relation of equilibrium contact angle to liquid and solid constitution. In: Fowkes FM (ed) Contact angle, wettability, and adhesion. American Chemical Society, Washington, D.C., pp 1–51
31. Boksanyi L, Liardon O, Kovats E (1976) Adv Colloid Interface Sci 6:95–137
32. Erard J-F, Nagy L, Kovats E (1983) Colloid Surf 9:109–132
33. Gobet J, Kovats E (1984) Adsorpt Sci Technol 1:111–122
34. Korösi G, Kovats E (1981) J Colloid Surf 2:315–355
35. Riedo F, Czencz M, Liardon O, Kovats E (1978) Helv Chim Acta 61:1912–1941
36. de Gennes PG (1985) Rev Mod Phys 57:827–863

37. Fowkes FM (1964) Contact angle, wettability, and adhesion. Advances in Chemistry Series, vol 43. American Chemical Society
38. Holloway PJ (1970) Pestic Sci 1:156–163
39. Öner D, McCarthy TJ (2000) Langmuir 16:7777–7782
40. Youngblood JP, McCarthy TJ (1999) Macromolecules 32:6800–6806
41. Walliser A (1992) Caracterisation des interactions liquide-fibres élémentaires par mouillage. Université de Haute Alsace (F), 92-MUHL-0248
42. Connor M (1995) Consolidation mechanisms and interfacial phenomena in thermoplastic powder impregnated composites. EPFL Thesis No 1413
43. Ruiz L (1997) Thesis: synthesis and characterisation of phosphorylcholine containing polymers designed to promote specific cell attachment via surface derivatisation. École Polytechnique Fédérale de Lausanne: Lausanne
44. Marieb EN (1998) Human anatomy and physiology, 4th edn. Addison-Wesley Publishing
45. Williams DF (1987) Definitions in biomaterials, vol 4. Progress in biomedical engineering. Elsevier
46. Kadoma Y, Nakabayashi N, Masuhara E, Yamauchi J (1978) Kobunshi Ronbunshu 35:423–427
47. Chapman D (1993) Langmuir 9:39–45
48. Hayward JA, Chapman D (1984) Biomaterials 5:135–142
49. Ishihara K, Aragaki R, Ueda T, Watenabe A, Nakabayashi N (1990) J Biomed Mater Res 24:1069–1077
50. Ishihara K, Aragaki R, Yamazaki JI, Ueda T, Watenabe A, Nakabayashi N (1990) Seitai Zairyo 8:231–236
51. Ishihara K, Oshida H, Endo Y, Ueda T, Watenabe A, Nakabayashi N (1992) J Biomed Mater Res 26:1543–1552
52. Lelah MD, Cooper SL (1986) Polyurethanes in medicine. CRC Press
53. Ruiz L, Johnston DS, Makohliso SA, Aebischer P, Mathieu HJ (1995) Biomimetic coatings on silicon wafers: synthesis and characterisation in ECASIA 95 Montreux, Switzerland. Mathieu HJ, Reike B, Briggs D (eds.) Wiley, Chichester
54. Yung LL, Cooper LS (1998) Biomaterials 19:31
55. Baumgartner JN, Yang CZ, Cooper SL (1997) Biomaterials 18:831
56. Ruiz L, Fine E, Vörös J, Makohliso SA, Léonard D, Johnston DS, Textor M, Mathieu HJ (1999) J Biomater Sci Polym E 10:931–955
57. Peyser P (1989) Glass transition temperatures of polymers. In: Brandrup J, Immergut EH (eds) Polymer handbook. Wiley, pp VI 209–277
58. Ruiz L, Hilborn JG, Léonard D, Mathieu HJ (1998) Biomaterials 19:987–998
59. Davies MC, Lynn RAP (1990) Crit Rev Biocompat 5:297–341
60. Yianni YP (1992) Biocompatible surfaces based upon biomembrane mimicry In: Quinn PJ, Cherry RJ (eds) Structural and dynamic properties of lipids and membranes. Portland Press, London, UK, pp 187–216
61. Hearn MJ, Briggs D (1988) Surf Interface Anal 11:198–213
62. Phillips MC, Finer EG, Hauser H (1972) Biochim Biophys Acta 290:397–402
63. Hauser H (1975) Phospholipid model membranes: demonstration of a structure -activity relationship in chemoreception. Infr Retr Ltd, London
64. Briggs D, Hearn MJ (1988) Surf Interface Anal 13:647–669
65. Chilkoti A, Castner DG, Ratner BD, Briggs D (1990) J Vac Sci Technol A8:2274
66. Tanuma S, Powell CJ, Penn DR (1993) Surf Interface Anal 21:165–176
67. Seah MP, Dench WA (1979) Surf Interface Anal 1:2–11
68. Garbassi F, Morra M, Occhiello E (1994) Surface energetics and contact angle. In: Polymer surfaces from physics to technology. Wiley, Chichester, chap 4, pp 161–219
69. Holly FJ, Refojo MF (1975) J Biomed Mater Res 9:315–326
70. Lavielle L, Schultz J (1985) J Colloid Interface Sci 106:438–445
71. Lavielle L, Schultz J, Sanfeld A (1985) J Colloid Interface Sci 106:446–451

72. Andrade JD (1988) Polymer surface dynamics. Plenum Press, New York London
73. Ratner BD, Weathersby P, Hoffman AS, Kelly MA, Scharpen LH (1978) J Appl Polym Sci 22:643–664
74. Wirpsza Z (1993) Polyurethanes: chemistry, technology and applications. Polymer science series. Ellis Horwood PTR Prentice Hall
75. Chevolot Y (1999) Thesis: Surface photoimmobilisation of aryl dyazirine containing carbohydrates – tools toward surface glycoengineering. École Polytechnique Fédérale de Lausanne, Lausanne
76. Lee YC, Lee RT (1995) Acc Chem Res 28:321–327
77. Monsigny M (1995) Biofutur 142:27–32
78. Petrak K (1994) Adv Drug Deliver Rev 13:211–213
79. Varki A (1995) Glycobiology 3:97–130
80. Hatanaka K, Takeshige H, Akaike T (1994) Carbohydr Chem 13:603–610
81. Hermanson GT, Mallia AK, Smith PK (1992) Immobilized affinity ligand techniques. Academic Press
82. Kobayashi K, Kobayashi A, Akaike T (1994) Method Enzymol 247:409–419
83. Onyiriuka EC (1990) Appl Spectrosc 44:808–811
84. Kobyashi K, Akaike T, Usui T (1994) Method Enzymol 242:226–235
85. Guire PE (1990) Method of improving the biocompatibility of solid surfaces. Pat 4–973–493, Bio-Metric Systems, Inc, US
86. Guire PE (1990) Biocompatible Coating for Solid Surfaces. Pat 4–979–959, Bio-Metric Systems, Inc, US
87. Erdtmann M, Keller R, Baumann H (1994) Biomaterials 15:1043–1049
88. Anderson AB, Tran TH, Hamilton MJ, Chudzik SJ, Hasting BP, Melchior MJ, Hergenrother RW (1996) Am J Neuroradiol 17:859–863
89. Anderson AB, Enrico LS, Melchior MJ, Pietig JA, Tran LV, Tran TH, Duquette PH (1994) Photochemical immobilization of heparin to reduce thrombogenesis. Tewntieth Annual Meeting of the Society for Biomaterials, Boston, MA, USA
90. Sigrist H, Collioud A, Clémence J-F, Gao H, Luginbühl R, Sänger M, Sundarababu G (1995) Opt Eng 34:2339–2347
91. Gao H, Sanger M, Luginbühl R, Sigrist H (1995) Biosens Bioelectron 10:317–328
92. Gao H, Luginbühl R, Sigrist H (1997) Sens Actuators B 38/39:38–41
93. Collioud A, Clémence J-F, Sänger M, Sigrist H (1993) Bioconjugate Chem 4:528- 536
94. Léonard D, Chevolot Y, Bucher O, Haenni W, Sigrist H, Mathieu HJ (1998) Surf Interface Anal 26:783–792
95. Chevolot Y, Bucher O, Léonard D, Mathieu HJ, Sigrist H (1999) Bioconjugate Chem 10:169–175
96. Léonard D, Chevolot Y, Bucher O, Sigrist H, Mathieu HJ (1998) Surf Interface Anal 26:793–799
97. Léonard D, Chevolot Y, Heger F, Martins J, Crout DHG, Sigrist H, Mathieu HJ (2001) Surf Interface Anal 2001:457–464
98. Yamashita K (1989) Method Enzymol 179:331
99. Stockert RJ, Morell AG, Ashwell G (1991) 12:441
100. Baenziger JU, Maynard Y (1980) J Biol Chem 255:4607–4613
101. Schwartz AL (1990) Annu Rev Immunol 8:195–229
102. Ashwell G, Harford J (1982) Annu Rev Biochem 51:531–554
103. Kawasaki T, Ashwell G (1976) J Biol Chem 251:12
104. Malissard M, Zeng S, Berger EG (1999) Bioconjugate Chem 16:125
105. Chevolot Y, Martins J, Milosevic N, Léonard D, Zeng S, Malissard M, Berger EG, Maier P, Mathieu HJ, Crout DHG, Sigrist H (2001) Bioorgan Med Chem 9:2943–2953
106. Sigrist H, Chevolot Y, Crout D, Martins J, Mathieu HJ, Lohmann D (2000) European Patent Application 99112422.3–2110

Plasma and Radiation-Induced Graft Modification of Polymers for Biomedical Applications

Bhuvanesh Gupta · Nishat Anjum

Departments of Textile Technology, Indian Institute of Technology, New Delhi-110016, India.
E-mail: bgupta@textile.iitd.ernet.in

Polymers have generated considerable interest in a number of innovative domains for biomedical applications. There is a great need to develop and design polymers for better acceptability to the biosystem that includes the blood and tissue compatibility and a perfect cell-material interaction. The modification of existing polymers by careful designing of the macromolecular architecture is one of the most attractive ways to develop such biomaterials. Plasma and high-energy radiation, such as gamma radiation and electron beam lead to enough activation of polymers so that specific monomer may be grafted. The extent of modification, i.e., the degree of grafting may be easily controlled by the careful variation of the radiation exposure and reaction conditions. The graft modification may be carried out almost on all polymers with the creation of a desired functional chemistry both on the surface and in the bulk matrix. Such biomaterials are being used for implants, sutures, tissue engineering, and as structured surface materials. The most recent innovation in the development of biomaterials has been in the area of tissue engineering where human cells are seeded on top of a polymer support and are harvested as tissue. The grafting method is extremely attractive as the modified biomaterial is obtained in the purest form possible. The grafting is applicable to almost all polymer-monomer combinations with enormous possibilities of physico-chemical and biological characteristics.

Keywords: Polymer, Biomaterial, Radiation, Plasma, Graft Polymerization, Modification

1	**Introduction**	36
2	**Graft Modification of Polymers**.	37
2.1	Plasma Modification	38
2.2	Ultra Violet (UV) Modification	42
2.3	Radiation-Induced Modification	43
3	**Applications**	48
3.1	Surface Construction	48
3.2	Sutures	51
3.3	Implants and Drug Release Systems	53
3.4	Tissue Engineering	55

4   Conclusion 58

References 59

1
Introduction

The use of polymers in medicine originated thousands of years ago when Egyptians were making use of fibrous materials as biomedical devices using naturally derived materials for wound closure. The scenario changed rapidly in this century with the birth of synthetic polymers having a wide range of physico-chemical characteristics. A large number of polymers available today have been investigated for their application as biomaterials, which has in fact led to the development of a few innovative materials finding places in medicine, surgery and pharmaceutical applications. Enormous efforts have been made to understand these materials and their application in the biosystem [1]. Biomaterials are materials which are derived from biological sources or which are used in contact with the human body and tissues such as implants or supplementary materials (vascular grafts, artificial hearts, sutures, and intraoccular lenses). Biomaterials require certain criteria for their performance in specific medical application, which takes into account the mechanical behavior, bioreceptivity, tissue compatibility, and immune as well as biological response. Commonly, polymers such as nylon, polyester, polypropylene, and Teflon are used as biomaterial. Since it is the surface of a biomaterial which first comes into contact with the living body when the material is placed in the biosystem, the performance of a material in biological environment would need a pertinent combination of surface characteristics and mechanical properties. The complexity arises because of the large differences in the requirements of a material due to the wide range of application sites in the human body. For example, a polymer material in contact with blood would experience cell adhesion, which subsequently could lead to thrombogenesis. At the same time, for polymers as implants, the tissue could provoke an adverse reaction in the form of inflammation.

The initial use of polymers in medicine was in the form of implants and wound dressings. The new era is witnessing a number of technically important domains such as tissue engineering where, based on the understanding of bio-receptivity and polymer-cell interaction, harvesting of an organ is achieved. This requires perfect knowledge of the biological surface and of material engineering, which opens up a challenging domain of human organ reconstruction and design, driving an expansion of the polymeric biomaterials field.

The cell-polymer interaction is believed to be mainly dependent on the physico-chemical properties of the material surface, surface free energy, hydrophilicity, and surface morphology [2–6]. The common polymers generally do not possess proper surfaces as required for biomedical applications. A functional biomaterial

Table 1. Physical and chemical surface modification methods [61]

Covalently attached coatings
Radiation grafting (Electron accelerator and Gamma)
Photografting (UV and visible sources)
Plasma (gas discharge) (RF, microwave, acoustic)
Gas phase deposition
Ion beam sputtering
Chemical vapor deposition
Chemical grafting (e.g., ozonization+grafting)
Silanization
Biological modification (biomolecule immobilization)
Modification of the original surface
Ion beam etching (e.g., argon, xenon)
Ion beam implantation (e.g., nitrogen)
Plasma etching (e.g., nitrogen, argon, oxygen, water vapor)
Corona discharge (in air)
Electron beam treatment
Ion exchange
UV irradiation
Chemical reaction
Non-specific oxidation (e.g., ozone)
Functional group modifications (oxidation, reduction)
Addition reactions (e.g., acetylation, chlorination)
Conversion coatings (phosphating, anodization)

would need its proper surface chemistry with excellent retention of bulk characteristics [7–9]. Both blood and tissue compatible materials have been modified by different methods involving the chemical and radiation routes. A comprehensive review of different modification methods is presented in Table 1. The table shows that quite different covalent and chemical modification methods exist. The covalent method offers a more stable modified surface as compared to other methods. The graft modification is one of such methods that provide covalent modification of materials with the creation of interesting architecture. Here the discussion will be confined to plasma and radiation-induced graft modification of polymers to produce biomaterials with pre-designed surfaces and bulk structures.

2
Graft Modification of Polymers

As outlined above it is desirable to modify polymers selectively for specific applications without losing much in terms of the inherent characteristics. The better the retention of the bulk properties, the more appropriate is the modification approach. In this respect, graft copolymers offer novel materials where the inherent polymer is represented by the backbone and the branches are formed by the graft-

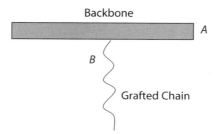

Backbone

A

B

Grafted Chain

Fig. 1. Schematic representation of the graft copolymer structure

ed monomer (Fig. 1). It is the grafted branches that introduce desirable properties in the polymer materials. If the polymer has hydrophobic character, the grafting of a polar monomer would lead to interaction between the material and water. This method is so promising that even a PTFE surface may be made into a hydrogel layer. Although there are different ways to synthesize graft copolymers, plasma and radiation methods have become the most versatile in terms of the physico-chemical behavior of the modified materials. These methods provide materials with high level of purity; they are surface-selective or may lead to the bulk modification of the matrix as well. The modification involves treatment of the polymer surface with plasma or high energy radiation to activate the surface or the bulk, respectively. This activated surface is subsequently treated with a polymerizable monomer under appropriate conditions so that the monomer is grafted on the polymer, thereby introducing desirable properties in the material based on the chemical nature of the monomer. The characteristics of the biomaterial would be therefore a function of the degree of grafting and graft distribution within the polymer or on its surface.

2.1
Plasma Modification

Plasma modification of polymers is an extremely useful way of tailoring a polymer into desired material by selective creation of the chemistry and molecular structure on the surface [10–12]. Plasma is the ionized state of a gas with an energy distribution in the range of 10–20 eV and is effective in controlled changes on the polymer surface. The surface chemistry may be altered by proper selection of the nature of the gaseous medium. The gases such as oxygen, ammonia, and carbon dioxide produce functionalities such as hydroperoxide, amino, and carboxylic groups, respectively. However, inert gases such as argon lead to the formation of radical sites on the polymer backbone, which are transformed into polar functionality in the presence of oxygen [13]. These functionalities act as the *'anchorage sites'* for the attachment of the biological molecules. In the former case, if the presence

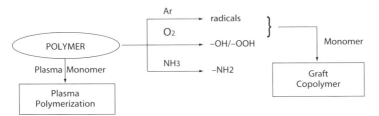

Fig. 2. Plasma modification of polymers

of the functionality is sufficient, the surface may become "bioreceptive". This has been observed in polyethylene films treated under ammonia or nitrogen/hydrogen plasma, which leads to the amination of the surface for the subsequent immobilization of protein via covalent coupling of glutaldehyde [14, 15]. The biomaterial scientists and technologists have been following this method successfully either by using the direct or the indirect method. In the first case, the monomer is exposed to very mild plasma environment (plasma deposition or plasma polymerization) in the presence of a polymer surface. In the second method, the polymer is exposed to the plasma and the activated surface is treated with a monomer, which initiates the graft polymerization (plasma grafting) (Fig. 2).

In the recent past, a wide range of monomers including both the aliphatic and fluorinated ones have been utilized for the polymerization where the extent of modification is governed by the plasma exposure conditions as well as by the chemical nature of the monomer [16]. The surface properties are altered by using monomers in the vapor phase for the plasma exposure. Allylamine and acrylic acid are polymerized on polyethylene (LDPE) surfaces where a brownish yellow film is formed [17]. However, the plasma polymerization of acrylic acid leads to transparent and highly crosslinked films on the surface and shows a less thrombogenic nature in terms of platelet adhesion and platelet spreading than the untreated PE surface. The surfaces developed at low plasma energy show higher thrombo-resistance. It is important to mention that during the process of surface construction in plasma polymerization, the monomer forms some sort of polymer coating on the surface. However, in graft polymerization a definite thickness on the surface is modified with significant alteration in the surface morphology depending on the degree of grafting. This is the reason that most of the researches on plasma modification in the recent time have gone to the tailoring of polymer surfaces by graft polymerization of different monomers [18–22].

PET surface may also be modified by plasma-induced graft polymerization of acrylic acid [18]. The exposure of the film under oxygen plasma leads to the formation of peroxide and hydroperoxide groups (Fig. 3). The modified surface contains a host of such polar functionalities and initiates the grafting of acrylic acid by

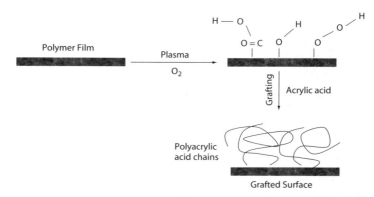

Fig. 3. Plasma-induced graft polymerization of acrylic acid into polymer films

decomposition of hydroperoxides at elevated temperature. The degree of grafting is considerably influenced by the plasma exposure parameters, such as plasma power, pressure, exposure time, and the reaction conditions involving monomer concentration and reaction time. The monomer concentration has profound influence on the degree of grafting (Fig. 4). The grafting increases with the increase in the acrylic acid concentration, reaches maximum, and then tends to decrease. The trend in grafting represents typical of a gel effect probably due to the viscosity enhancement and the lack of termination of growing chains. At higher monomer concentration, the extensive homopolymerization may proceed leaving behind very little monomer for the grafting reaction. Similar trend in the N-vinylpyrrolidone grafting on polysulfone has been observed where a continuous increase in the graft levels in the whole range of monomer concentration (up to ~10%) takes place [23]. This approach has been extended to the grafting on silicone rubber by exposing the Ar plasma treated surface to oxygen atmosphere [24]. This hydroperoxidized surface is subsequently treated with 2-hydroxyethyl methacrylate (HEMA) to introduce hydroxyl functionality. It is observed that the plasma treatment time influences the graft add-on, significantly (Fig. 5). There is an optimum plasma treatment time that is required for the maximum grafting to take place. The higher exposure may lead to the loss of active centers that are otherwise responsible for the peroxidation during exposure to oxygen – an observation made for the grafting of acrylamide onto polyethylene surfaces as well [25].

Modification of polyurethanes as a blood contacting material is achieved by coating the polymer surface with heparin as the most suitable biomolecule that avoids thrombogenesis. The immobilization needs a functionalized surface that may offer sites for the heparin bonding. It is observed that helium plasma treatment of polyacrylonitrile and polysulfone followed by exposure to air leads to peroxide species of the order of 10 nmol/cm^2, which are available for decomposition

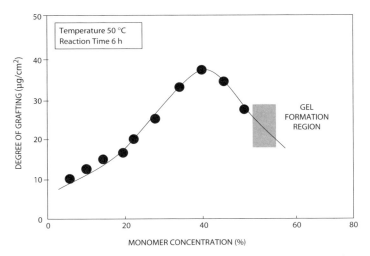

Fig. 4. Variation of the degree of grafting with the acrylic acid concentration on plasma treated PET films [18]

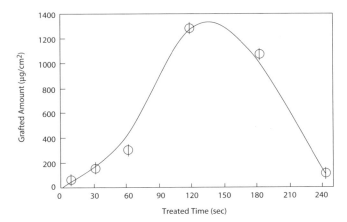

Fig. 5. The effect of the amounts of grafted PHEMA on Ar-plasma treated from 0 to 240 s of treated time (60 W, 200 mtorr) [24]

to initiate the grafting process [26]. Although, the magnitude of peroxidation would depend on the nature of the polymer, the oxidative functionality is sufficient for the initiation of the grafting of acrylic acid and methyl acrylate on the polyurethane surface [27–29]. Since this way a sufficient amount of carboxyl groups is fixed at the surface, these groups may be directly used for the reaction with heparin and insulin. The carboxyl groups offer sites for the coupling with amino group terminated polyethylene oxide (PEO) where they are directly used for the interaction with heparin by carbodiimide activation method.

PEO is a non-toxic water-soluble polymer which is extensively used for biomedical applications. PEO resists the adsorption of plasma proteins because of their strong hydrophilicity, chain mobility, and lack of ionic charge [30]. The PEO immobilization on any polymer surface would result in the decrease in the protein adsorption and platelet adhesion. This versatile nature of PEO has led to many studies not only on the preparation of PEO derived surfaces but also on the subsequent uses of these surfaces as biomaterials [31–34]. The PEO layer if adhered on the normal material surface may be washed off because of the absence of any interaction between the two surfaces. Therefore, covalent immobilization of PEO has been carried out either by grafting PEO molecule on a base polymer or by grafting a monomer bearing pendant PEO groups.

Fluorinated surfaces are known to have low surface energy and any attempt to fluorinate a polymer surface would lead to a biomaterial with low profile in protein adsorption. Vinylidene fluoride and chlorotrifluoroethylene have been grafted onto polyethylene so that the polymer surface acquires higher hydrophobicity as evident from the enhancement in the contact angles of the modified films. An alternative to the development of a heparinized polyurethane surface is the treatment of an oxygen-plasma-exposed-surface with 1-acryloylbenzotriazole. The grafted surface is hydrolyzed to carboxylic groups or is subsequently aminated to free amine so that heparin may be immobilized by the carbodimide route [35].

2.2
Ultra Violet (UV) Modification

UV irradiation of polymers has been observed to be an effective technique to modify polymers for biomolecule immobilization. The polymer surface is modified by photo-induced graft polymerization of different monomers, such as acrylic acid, 2-acrylamidomethylpropane sulfonic acid, and styrene sulfonic acid. All these monomers have ionic functionality and lead to high graft densities for the immobilization of collagen and other proteins. The surface density of polyacrylic acid grafts reaches up to $100\ \mu g/cm^2$ under appropriate reaction conditions and is enough for protein immobilization [36].

The hydrophilicity of the catheter surface is an important criterion for enhancing lubricity of the biomaterial and has been introduced by UV induced grafting of polar monomers [37]. In an attempt to produce balloon catheters, polyethylene films and catheters are grafted with dimethylaminoethyl methacrylate (DMAEMA) and acrylic acid under UV irradiation [38]. The acrylic acid leads to much higher graft density as compared to the acrylate monomer and an autoacceleration in the grafting of acrylic acid is observed (Fig. 6). The grafting of DMAEMA, on the other hand, does not show any acceleration and levels off much faster. In an-

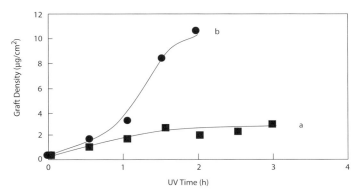

Fig. 6. Effect of UV irradiation time on grafting density using (*filled squares*) DMAEMA 1 wt% and (*filled circles*) AAc 10 wt% solution [38]

other attempt, the grafting of glycidyl methacrylate on polymers has been carried out with a view that the epoxy group offers a versatile site for the covalent immobilization of biomolecules [39]. The epoxy group opens up in the presence of different functional groups, such as amino, carboxyl, and hydroperoxides and forms stable bonding with different molecules. In order to achieve high graft densities, degassing of monomers is an important step prior to the grafting. However, this may be overcome by the addition of some agents such as sodium periodate ($NaIO_4$), which helps in oxygen depletion thereby allowing the grafting reaction to proceed in an efficient manner [40].

Polymers such as styrene-butadiene-styrene block copolymers have been modified by graft polymerization of vinyl pyridine, which is quaternized with iodomethane. Such a membrane is then able to bind heparin as a function of the amount of the grafted component [41]. A combination of plasma and UV irradiation may be used for the surface grafting on polymer surfaces. Silicon surfaces are exposed to argon plasma with subsequent exposure to air in order to introduce peroxides and hydroperoxides. The oxidized surface is subsequently grafted with polyethyleneglycol methacrylate by UV irradiation [42]. Even a short plasma exposure of 10 s is enough to generate functionality for the UV grafting. As a general trend, the graft densities on silicon surfaces show an increasing trend with the increase in monomer concentration and UV exposure time.

2.3
Radiation-Induced Modification

As the third technique for polymer modification, radiation-induced graft polymerization has been extensively used since the starting material polymer not only acquires required properties but also retains most of its inherent characteristics.

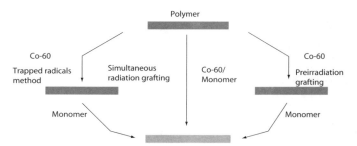

Fig. 7. Schematic representation of the radiation grafting process

Both gamma rays and electron beam are used for the activation of polymers. The radical sites are created uniformly across the bulk of the material so that the modification becomes uniform across the matrix. Moreover, the graft densities may be easily controlled by proper selection of the irradiation and synthesis conditions [43]. One of the major attractions of this modification is that the biomaterial thus produced is highly pure as no initiator and related impurities remain in the matrix. Several studies have been devoted to the development of biomaterials based on the radiation grafting process [44–46]. There are three distinct approaches for the graft polymerization of a monomer into a polymer, i.e., simultaneous irradiation, preirradiation in air, and trapped radicals methods (Fig. 7). Whatever method one adopts for the grafting reaction, the final product is almost identical but with some variation in the physico-chemical structure. Significantly high radiation doses are needed to achieve sufficient concentration of active species in the preirradiation method as compared to the simultaneous irradiation method. As a result, additional physical changes in the biomaterial matrix are imparted. It is important to note that the electron beam activation needs much shorter times as compared to the gamma irradiation and hence the degradation processes are better contained in electron beam irradiation.

Blood compatible materials need to be designed in such a way that thrombus formation (adsorption of certain plasma proteins and adhesion of platelets) is minimal. Since it is the surface of the material that induces thrombogenesis, proper tailoring of the surface would be beneficial in improving biocompatibility without altering the bulk properties of biomaterials. Polyethylene oxide (PEO) is a material with significantly high blood compatibility. As discussed earlier, the nonspecific protein adsorption on PEO surfaces is minimized because of the absence of any partial charges in spite of the increase in the surface energy. The low affinity of PEO for proteins and other blood components has stimulated many investigators to study the blood-PEO based biomaterial interaction [47, 48]. The PEO-biomaterial is obtained from a polyethylene precursor modified by graft polymerization

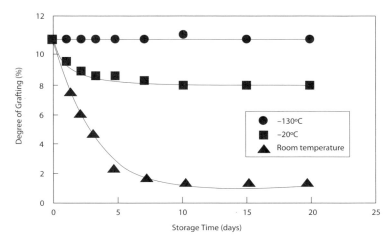

Fig. 8. Effect of storage temperature and time on the grafting of PEO4/5- MA onto 120 kGy preirradiated polyethylene in 30 vol.% PEG-MA MeOH/THF (2/1) solution at 30 °C for 3 h [49]

of polyethyleneglycol methacrylate by preirradiation method. The degree of grafting is found to be strongly dependent on the reaction conditions. As to be expected for a radical-initiated reaction, the storage time of the irradiated film prior to the reaction influences the degree of grafting considerably. The storage at lower temperature is very effective in retaining the grafting levels (Fig. 8). The room temperature storage, on the other hand, leads to drastic reduction in the grafting, while storage at −130 °C produces no change in the grafting even after 20 days of storage [49]. It seems that at that temperature the segmental mobility is so low that the free radicals that are produced during the irradiation remain trapped within the matrix. It is important to mention that the initial loss in radical concentration and hence the grafting ability depends on the nature of polymers. While low-T_g hydrocarbon polymers such as polyethylene show some loss, the high-T_g fluoropolymers such as tetrafluoroethylene-*co*-hexafluoropropylene (FEP) do not undergo a strong decrease in grafting ability even after the storage for three months [50, 51].

Radiation-induced cografting of polyethyleneglycol methacrylate (PEGMA) and methoxypolyethyleneglycol methacrylate (MPEGMA) with HEMA onto silastic films has been carried to produce non-fouling polymer surfaces [52]. The extent of grafting is influenced by the monomer concentration and the relative composition of the two monomers. An increase in the PEG monomers in the mixture leads to the decrease in the grafting because of the low reactivity of the PEG monomers as compared to HEMA (Fig. 9). Virtually little grafting proceeds in the medium comprising pure PEG monomers. It is important to see that the molecular weight of the PEG unit in the monomer influences the degree of grafting signifi-

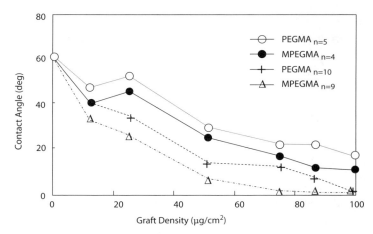

Fig. 9. Grafting yield of Silastic film as a function of PEGMA/or MPEGMA mol % in comonomer: [HEMA+PEGMA/or MPEGMA]=0.4 mol/l, [Cu(NO$_3$)$_2$]=0.1; dose: 0.63 Mrads; solvent; MeOH-H$_2$0 (20:80 vol/vol) [52]

cantly. The grafting yield decreases with the increase in the PEG molecular weight in a monomer. This decrease is confined to an 'n' value of 20, beyond which a plateau in the graft level is attained. The decrease in grafting is associated with a steric effect of the long polymer chain. The shielding by PEG group may also lead to the lower graft yields.

A morphological investigation on the biomaterial surface shows that, at low graft level, the grafts are confined more to the surface leaving behind the bulk more or less intact. Once the grafting increases on the surface, the thickness of grafted layer also increases. This is due to the fact that the grafting proceeds by a '*diffusion controlled*' mechanism where initial grafting takes place at the surface only and proceeds further by progressive diffusion of monomer through the medium swollen grafted layers [50]. It has been observed that grafted monomer forms a distinct phase within the polymer matrix and may lead to the incompatibility and surface nonhomogenity depending on the extent of grafting.

The grafting of *non*-PEG monomers such as acrylamide and HEMA has also been carried out on polymers to introduce antithrombogenic characteristics [53, 54]. The degree of grafting is considerably influenced by the irradiation conditions, film thickness, and soaking time. It is observed that the nature of the solvent plays an important role in the grafting process [53]. Of all solvents such as water, benzene, dichloroethane, and chlorobenzene, water showed the least grafting because of its nonswelling nature towards PP (Fig. 10). All other solvents act as swelling agents for PP and result in maximum possible area for the monomer diffusion into the polymer matrix so that additional grafting takes place in the polymer bulk.

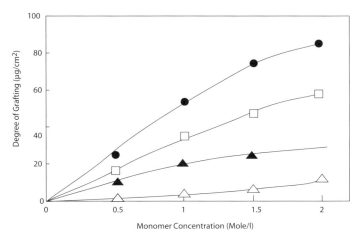

Fig. 10. Variation of percent grafting with monomer concentration in different solvents: (*open triangles*) water, (*filled triangles*) dichloroethane, (*open squares*) benzene, (*filled circles*) chlorobenzene; dose rate, 56 rad/s; dose, 0.3 Mrad [53]

Similar behavior is observed in the grafting of HEMA on chitosan where solvent composition influences the graft levels in such a way that a maximum is observed at 50% in water-methanol mixture [54]. Probably water addition to the grafting mixture helps in better accessibility of the monomer to the grafting sites within the bulk. Once methanol content in the mixture increases beyond 50%, the grafting decreases due to the inhibitory action of methanol. Similar observations have been made in the acrylamide grafting into polyethylene fibers [55]. Radiation grafting of acrylic acid onto polypropylene has been carried out to develop antimicrobial materials. It is found that the addition of organic and inorganic chemicals to the grafting reaction influences the degree of grafting considerably [56]. Among the acids, sulfuric acid addition enhances the grafting almost two-fold while acetic acid has the least influence. However, the addition of inorganic salts such as ferrous sulfate and copper sulfate diminishes the degree of grafting due to the inactivation of growing chains by metal ions. Attempts have been made to modify polymer surfaces by grafting thermosensitive polymers. Such a biomaterial may be expected to provide thermo-responsive conformational changes in the body. In this process N-isopropylacrylamide has been grafted on to polystyrene surface [57]. The electron beam grafting led to a surface with 1.9–2 $\mu g/cm^2$ graft with polyisopropylacrylamide.

3
Applications

3.1
Surface Construction

The initial response of the body towards a biomaterial depends on the surface characteristics because it is the surface of the biomaterial that comes in contact with the living tissues. Hence, proper designing of the surface is of utmost importance for enhanced compatibility of the biomaterial. All the techniques discussed above produce biomaterials with desired surfaces depending on the chemical nature of the polymer. The magnitude of the surface changes is reflected in contact angle variation of polymers as a result of specific modification. The higher the hydrophilicity of the surface, the higher is the decrease in the contact angle. This is reflected in the water contact angle and surface energy of the biomaterial as the graft level increases [39, 41]. Acrylic acid (AA_c) grafting leads to much higher decrease in contact angle as compared to the 2-dimethyaminoethyl methacrylate on PE surface [38]. It is observed that the ionization of the functional groups leads to the further lowering in contact angle and reaches a value less than 20° for acrylic acid grafted PE surface (Fig. 11). The trend in the contact angle variation in fact depends more on the hydrophilicity of the monomer; the nature of the polymer has little impact on that. This is what has been observed in the identical trends in

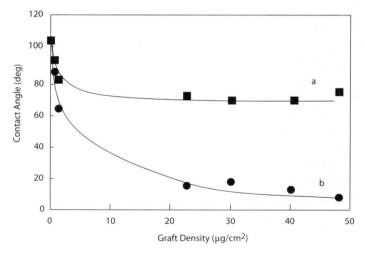

Fig. 11. Contact angle dependence on graft density of AAc-grafted PE films: *filled squares* grafted film, *filled circles* grafted films immersed in 0.1 N NaOH for 45 min [38]

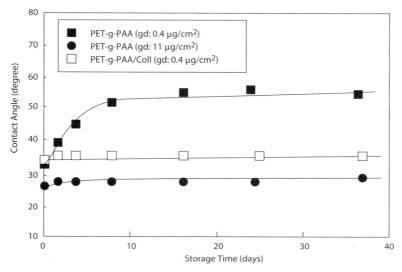

Fig. 12. Variation of contact angle of plasma grafted PET films before and after collagen immobilization

contact angle variations in acrylic acid and acrylamide grafting on FEP, PE, and PET surfaces [38, 58–60].

The surface morphology of materials is often considerably influenced by the graft modification irrespective of the type of modification [52, 59]. The surface roughness is more severe after plasma treatment due to the etching process as compared to the gamma radiation grafting. Plasma leads to a hill-valley structure on the surface with a nano-scale frequency [13]. As a matter of fact, surface roughness increases with the increase in the graft levels in a polymer. At higher graft levels the grafted moiety forms independent domains within the polymer matrix. As a result, polymer chains are pushed apart leading to a non-homogeneity on the surface. However, any post-grafting immobilization would further alter the surface morphology as observed in the collagen immobilization. Collagen being a macromolecule forms independent structures on the surface and overshadows the inherent non-homogeneity on the grafted surface. It is therefore the hydrophilicity and contact angle of the collagen immobilized surface that needs to be considered for subsequent applications (Fig. 12).

The most critical requirement in the surface design is the development of a stable surface. The surface undergoes rearrangement of atoms and molecules in response to the external environment and leads to changes in the chemistry and morphology [61]. The problem of surface instability is more severe with the plasma exposed materials where the modification is confined to a very thin layer of few

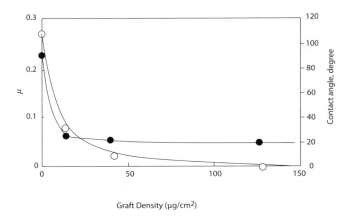

Fig. 13. Coefficient of friction (μ) (*open circles*) and contact angle (*filled circles*) of acrylamide-grafted polypropylene film as a function of graft density [7]

nanometers and polymer chains tend to minimize the surface energy by movement of this functionality towards the bulk of the matrix [18]. The surface energy may be altered by linking these surface functional groups with a large molecule by graft polymerization. Graft polymerization of acrylic acid onto oxygen plasma exposed PET surface has been effective in inhibiting the surface migration. The polyacrylic acid grafts, as a result of bulky structures, reduce the mobility of surface chains and hinder the surface rearrangement. This is evident from the contact angle variation of the grafted surfaces, which remains almost identical for the whole range of the storage periods.

Tubular devices, such as catheters, cannulae, and endoscopes are inserted in the body orifices. Effective lubrication of the device surface is needed to overcome the frictional resistance between the device surface and the mucous membrane during insertion or removal. The coefficient of friction is usually a function of several variables including material, configuration, and applied tension. Thus braided sutures have higher friction as compared to the monofilament ones. The graft polymerization of hydrophilic monomers onto device surface is extremely useful in modifying these surfaces [62, 63]. The grafted device surface by virtue of the hydrogel nature absorbs water and undergoes quick swelling, which leads to significant reduction in the coefficient of friction. The grafting of dimethylacrylamide onto polypropylene surface leads to a material with high level of hydrophilicity, which makes the surface slippery. As a result, the coefficient of friction decreases fast (Fig. 13).

3.2
Sutures

Considerable work has been carried out on the design of surgical sutures and prosthetic devices by grafting specific monomers onto a suture surface. Hydrogels are good candidates for biomedical applications because of their excellent swelling in aqueous media, which provides a continuous path for permeation and diffusion of low molecular weight metabolites in tissues. Hydrogels show very weak mechanical characteristics in the swollen state and therefore their application as suture material is restricted. One of the ways to have hydrogels for biomedical application is to graft them on polymer supports such as PET, Nylon, and PP monofilaments and multifilaments. As a result, one can achieve a surface where bulk properties of the polymer would be retained to a large extent and the hydrogel surface would perform the job of a biocompatible layer. Several studies have been carried out on the radiation-induced grafting of such monomers as 2-hydroxyethylmethacrylate (HEMA), acrylic acid, acrylamide, and N-vinylpyrrolidone onto inert polymers to produce a material with the surface properties of a hydrogel and with better mechanical properties [64–66]. The advantage of the grafting method over the modification through additives is that in the latter case incompatibilities and phase separation persist which adversely affect the physical properties of the sutures.

The development of antimicrobial sutures is a step further in the area of biomaterials. The main drawback of braided sutures is the growth of bacteria in the interspaces between filaments, which causes infection [67]. The grafting of polar or ionic monomers onto these braided sutures may lead to a surface with ionic functionality where a suitable drug may be immobilized. Antimicrobial silk sutures have been developed by the grafting of methacrylic acid (MAA) onto braided multifilament and subsequent immobilization of 8-hydroxyquinoline drug on the

Table 2. Antimicrobial activity of 8-HQ immobilized Silk-g-methacrylic acid samples (degree of grafting: 63%) [68]

Microorganism	Sample	Antimicrobial activity zone of inhibition (cm^2)	
		Before release	After 32 days release
E. coli	Pure silk	0	0
	Silk-g-MAA	0	0
	Silk-g-MAA+8-HQ (25.2%)	19.6	2.5
S. aureus	Pure silk	0	0
	Silk-g-MAA	0	0
	Silk-g-MAA+8-HQ (25.2%)	10.2	2.0
P. aeruginosa	Pure silk	0	0
	Silk-g-MAA	0	0
	Silk-g-MAA+8-HQ (25.2%)	7.1	0

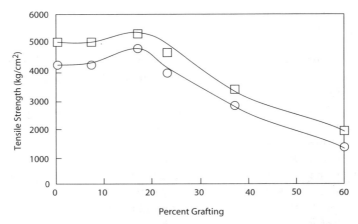

Fig. 14. Variation of tensile strength of polypropylene suture with percent grafting (*open squares*) straight and (*open circles*) knotted [69]

grafted surface [68]. As the degree of grafting increases, the drug loading on suture increases because of the increased interaction between the carboxyl groups and the drug. The antimicrobial activity of these sutures against various bacteria as measured by the zone of inhibition shows that the sutures are effective against microorganisms and possess effective antimicrobial characteristics (Table 2).

The graft polymerization of monomers on polymers is always accompanied by an alteration in the physical structure of the biomaterial. The most pronounced changes occur in the crystalline regions and have profound influence over the mechanical properties [69]. The crystallinity in gamma radiation grafted PP suture decreases to two third of the original value for a graft level of 60% polyHEMA. These changes need not be considered as a loss of crystalline zones of PP. Instead, a dilution of the inherent crystallinity takes place due to the incorporation of amorphous grafted moiety. The tensile strength also varies with the graft add-on. Tensile strength increases initially but a sharp reduction is observed beyond 17% graft add-on (Fig. 14). The initial increase arises from the filler action of polyHE-MA microstructure within the amorphous region of polypropylene. At higher graft add-on, the compactness of chains is adversely affected due to pushing apart of chains and leads to the loss in tensile strength. In order to overcome the incompatibility of an ionic monomer grafted suture, non-ionic monomers such as acrylonitrile are grafted onto a PP suture. The tensile retention in such a suture is much better than an acrylic acid grafted one. The nitrile group in the grafted suture is so versatile that it may be transformed to various functional groups such as amino, amide, and carboxylic sites. Another advantage of the acrylonitrile hydrolysis route over the acrylic acid one is that very little homopolymerization occurs during the reaction so that relatively pure grafts are produced without the impurity of the occluded homopolymer. Nylons have also been modified by graft polymerization of acrylic acid so that the carboxylic acid groups may be used to immobilize antibiotics such as penicillin and gentamycin which introduces the antibacterial property in relation to Gram positive and Gram negative bacteria [70]. The metallic complexes of polymeric material have been observed to extent antimicrobial behavior [71]. In such cases, the functionalized surface such as polypropylene-g-polyacrylic acid produced by preirradiation grafting is complexed with metallic salts. The complexation occurs with the carboxylic group and behaves as the antimicrobial material most probably by ionic interaction.

3.3
Implants and Drug Release Systems

There is an increasing demand for blood and tissue contacting materials in the medical domain. Coupling proteins and other biomolecules to a polymer surface is an attractive way to develop surfaces for medical applications. The haemocompatible materials require the least interaction of blood proteins and platelet adhesion. Since polyethylene oxide is known for the low affinity towards proteins, the grafting of polyethyleneglycol methacrylate on polyethylene film has been carried out to achieve this task [5]. This as a result leads to the reduced amount of adsorbed plasma protein as the degree of grafting increases. The platelet adhesion is also suppressed considerably on the grafted surfaces.

Heparin may also be immobilized onto the plasma grafted surfaces for the antithrombogenic characteristics. By virtue of its anticoagulant behavior, the modified polymer surface offers antithrombogenic properties. PET knitted fabric has its application in prosthetic arterial graft where it is necessary that the surface is blood compatible. This has been achieved by the grafting of acrylic acid on PET fabric followed by the coupling of PEO as the spacer [58]. The heparin is subsequently immobilized on this modified surface (Fig. 15). The thrombus formation on polyacrylic acid grafted PET does not show much variation as compared to the virgin PET. However, PEO-linked grafted surface shows significant reduction in the thrombus formation. The heparin immobilization further leads to reduced thrombus formation on the knittings. A similar approach has been followed towards the development of blood compatible polyurethane by graft polymerization of acrylic acid and methylacrylate [27]. Polyurethane is one of the most widely used biomaterials as catheter and heart devices due to its excellent mechanical properties and good blood tolerability. Its antithrombogenic characteristics may be further improved by incorporating heparin. The polyurethane surfaces having PEO content of 32–47 nmole/cm^2 lead to the heparin immobilization in the range of 1.16–1.3 µg/cm^2. Since the heparin is covalently bonded, the selectivity of the surface is much higher than the ionically immobilized heparin (Table 3).

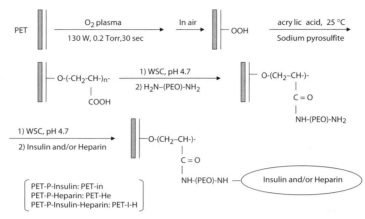

Fig. 15. Schematic diagram showing the immobilization of insulin and/or heparin on PETs [58]

Table 3. Concentration of PEO and heparin immobilized onto PU surfaces [27]

Sample	Concentration of PEO (nmole cm^{-2})	Amount of heparin (μg cm^{-2})[a]
PU-6	42±4	
PU-33	32±5	
PU-6-Hep	42±4	1.3±0.09
PU-33-Hep	32±5	1.16±0.07

[a]Measured with toluidine blue (n=4)

The heparin immobilized polyurethane surfaces (prepared by plasma grafting of 1-acryloyl benzotriazole and subsequent hydrolysis/amination) are effective in suppressing thrombus formation. The adhesion of peripheral blood mononuclear cells was also lower on such modified surfaces [72]. Heparin immobilized on vinyl pyridine grafted styrene-butadiene-styrene block copolymers also shows good biocompatibility. The adsorption of albumin and fibrinogen are reduced with the increasing graft levels and heparin content [41].

Thermo-sensitive polymers have been evaluated as well for platelet adhesion [73]. The platelet behavior on these grafted surfaces depends on the temperature. Below the lower critical solution temperature (LCST) of 32 °C in poly(N-isopropylacrylamide) grafted surface, the platelets on a grafted surface display a round shape and oscillating vibratory micro-Brownian motion. This observation is similar to the PEG-grafted surfaces where platelet activation is inhibited. Above LCST, surfaces facilitate the cell adhesion and culture where cells readily adhered, spread, and developed the characteristics of pseudopodia on the surface. The controlled delivery of drugs from modified polymers has been an interesting area. Polymers are modified by grafting appropriate monomer bearing polar or ion exchange sites. A suitable drug is immobilized on the surface to achieve the required dose. As soon as this drug loaded polymer comes into contact with the body, the matrix swells and it releases the drug to the external environment.

3.4
Tissue Engineering

Millions of surgical procedures are performed annually to treat patients suffering from organ disorders. With the increasing demand for polymers in tissue engineering in treating patients for the loss or failure of an organ or tissue, it is becoming necessary to design and develop a polymer for immobilization of biologically active molecules and living cells [73–77]. The polymers are made into scaffolds of required shape and size and living cells from the patient are seeded onto the polymer surface and are harvested as tissue for its subsequent transplantation to the patient. The expanded tissue is therefore structurally integrated in the body. This overcomes the problem of donor shortage and transplant rejection of the biomaterial [78]. The innovations in this area are that it opens up a straight forward route to the harvesting of many of the body parts without biological complications.

Tissue engineering presents enormous challenges and opportunities for material scientists from the perspective of both the material design and material processing. The polymers have either no functional groups or a very low level of functionality if at all, which may not necessarily lead to a bio-interactive surface. Such polymers, therefore, need selective modification of their surfaces to introduce functionality, such as hydroxyl, carboxyl, amino, and imino groups in a sufficiently large quantity. These functional groups act as sites for the immobilization of proteins and subsequently seeding of human cells.

The modification of polymers may be achieved by the graft polymerization of specific monomers by ultraviolet, plasma, and high energy radiation. The first two methods are surface selective and hence the graft modification remains confined to the depth of a few nanometers on the surface of the polymer. The plasma-induced graft modification has proven highly successful as a means to develop functional interfaces for the immobilization of biomolecules and cell cultures [79, 80]. The graft polymerization on the activated polymer leads to brush like branches on the polymer surface, which are involved in the protein interaction. PET surface in the form of films, fibers, and fabrics has been extensively studied for the immobilization of biomolecules on the grafted surface [81]. Collagen is the most interesting surface for tissue adhesion and its proliferation by direct attachment on the substrate surface. Hence, the collagen immobilization onto a polymer support would be the proper interface for extra cellular matrix (ECM)-tissue interaction. It is found that the grafting of acrylic acid leads to a surface that contains carboxyl

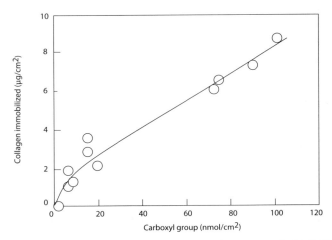

Fig. 16. Surface density of the collagen immobilized onto the PAAc-grafted PET films through interpolymer complexation [36]

functionality where collagen may be immobilized by the dip-coating process. The higher the carboxyl content, the higher is the collagen immobilization due to increased functional group interaction (Fig. 16). The slightly negatively charged polyacrylic acid, having some deprotonated carboxylic acid functions, interacts with the positively charged protonated amines on the collagen to form an ionically crosslinked surface. The amount of collagen on PET surface depends on the degree of grafting and relatively high amounts of collagen (>10 µg/cm²) could be immobilized on the surface (Fig. 17). The washing of the dip-coated surface, however, leads to almost 50% loss in collagen. This arises due to the loosely bound collagen being washed away.

Urinary bladder reconstruction is one of the most recent innovations in the domain of the tissue engineering. The bladder urothelial cells and smooth muscle cells may be seeded onto a polymer support, cultured in vitro, and expanded to a required size [82–86]. Here textile materials offer innovative surfaces for bladder construction [83]. Knitted scaffolds out of both the biostable scaffolds – such as polyester – and the biodegradable ones – such as polyglycolic acid – may be used for the cell seeding. The collagen immobilized PET surfaces provide excellent interface for the seeding of urothelial cells and smooth muscle cells. These cells adhere and grow well on the collagen immobilized surfaces (Fig. 18). The knitted PET structures offer a better simulated matrix for bladder construction. The advantage is that they elongate under stress. The large hysteresis of the knitted structures is not only due to the hysteresis of the yarn itself but mainly due to the straightening of the loops and the possibility of the movement of the yarn relative

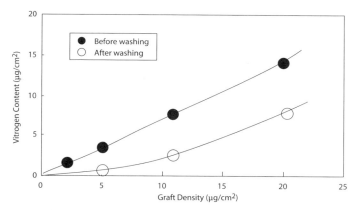

Fig. 17. Variation of Vitrogen content with the graft density on plasma grafted PET-g-PAA films [59]

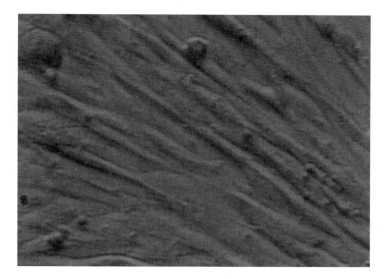

Fig. 18. Smooth muscle growth on Vitrogen immobilized PET-g-PAA film (5 µg/cm^2) PAA [59]

to the individual loops. The important aspect of knitted structures is that the deformation is nearly reversible. Hence, the growth of urothelial cells and smooth muscle cells on the knitted surfaces may lead to a more appropriate platform for the cell seeding [84–85]. Biomimetic surfaces may be produced by UV graft modification of PE surface with vinylmethoxysilane and subsequent hydrolysis [86].

The modified surface, as investigated in a simulated body fluid, showed a dense and homogeneous bone mineral like apatite layer even on fibers, which may be subsequently woven up into a composite structure to produce a three-dimensional structure analogous to that of natural bone.

There is an increasing demand for biodegradable polymers for scaffold preparation. Poly(α-hydroxy acids) such as polyglycolic acid and polylactic acid and their copolymers offer the required physico-chemical characteristics for their function as extra-cellular matrix in tissue engineering. These polymers may not necessarily need a graft modification to enhance the surface functionality; instead surface hydrolysis may be enough for a functionalized surface. The advantage of such polymers is that their degradation rate may be easily controlled by careful composition of the copolymers. The extra-cellular matrix degrades and leaves behind an intact patch of the tissue at the transplantation site.

Tissues engineering has been successful in designing and transplantation of artificial hair into human skull [36]. The artificial hair is fabricated from poly(ethylene terephthalate) monofilament where the tissue contacting part of the filament is modified to make it biocompatible. In this process, the monofilament is modified by grafting of acrylic acid using suitable initiation process, such as UV radiation. This modified filament is subsequently immobilized with collagen using glutaraldehyde crosslinker. Cells may be immobilized onto this modified surface for subsequent implantation into the human skull. This procedure is so exciting that it has been applied to more than 110,000 scalps for hair replacements world wide. Vascular grafts are also important domains of tissue engineering, as shown by approximately 750,000 coronary artery bypass graft surgeries. The traditional method of treatment is to use a blood vessel from another location in the patient (autograft), e.g., from the leg. The problem with the autograft procedure is that it requires multiple surgeries and increased risk and cost to the patient. Moreover 30% of bypass patients lack suitable vessels. Tissue engineered grafts therefore provide a feasible alternative. There are several factors to consider in graft development, including the permeability, since it is a fluid-containing device, and the flexibility, it must be readily handled and sutured securely into place. Braided fibers and knitted textiles are excellent materials for the ECM which may be modified by incorporating polar functional groups for collagen immobilization where cells may be seeded in a three-dimensional manner and a graft may be produced.

4
Conclusion

Polymers have developed enormous interest in a number of bio- and medical related applications where physico-chemical characteristics of both the surface and the bulk need to be tailored. Since the biomaterial remains in close contact with the living body, the reaction from the tissues would be a function of the surface characteristics. Almost no material can act as a completely biocompatible one. Therefore it is needed to develop polymeric materials with the minimum adverse response from the body. The polymer surfaces needs proper designing in terms of the chemical functionality in such a way that the compatibility of the biomaterial with the body is enhanced. Graft polymerization of specific monomers offers an attractive route towards the modification of polymers so that the material acquires desired properties while retaining most of the inherent characteristics. The extent of compatibility, therefore, rests with the chemical nature of the monomer and its content in the modified matrix.

Both plasma and high energy radiation are being used for the selective introduction of chemical functionality. Plasma-induced graft polymerization is an innovative way in terms of the nano-scale thickness modification of the surface. This is an advantage in terms of almost complete retention of the mechanical strength of the biomaterial after graft modification. This is helpful in designing surfaces for innovative areas of tissue engineering where human cells are seeded on the nano-structured surface for subsequent harvesting into a tissue. The radiation, on the other hand, leads to surface and bulk modification of the material. As the grafted component remains thoroughly distributed across the matrix, the whole matrix is functionalized but the tensile strength may be adversely affected. It is important to mention here that the tensile strength will be a function of the amount of the grafted component and its compatibility with the original matrix. There is a wide spectrum of application areas of biomaterials. The area of tissue engineering has further opened up a newer domain of biomaterials where proper simulation of polymer-cell interaction needs to be followed.

As a matter of fact the role of polymers as biomaterial is spreading fast. Many more domains should be coming up in the next generation of materials where more attention on the matching of the physical properties with the chemical functionalization will be required. Here, proper weaving of the knowledge of a material scientist and a biotechnologist will be required to upgrade the concept of biomaterial in innovative domains.

References

1. Wise DL, Trantolo DJ, Altobelli DE, Yaszemski MJ, Gresser JD, Schwartz, ER (1995) Encyclopedic handbook of biomaterials and bioengineering. Marcel Dekker, New York
2. Van Kampen CL (1979) J Biomed Mater Res 13:517
3. Absolom DR, Thomson C, Hawthorn LA, Zingg W, Neumann AW (1988) J Biomed Mater Res 22:215
4. Ratner BD, Hoffman AS, Hanson SR, Harker LA, Whiffen JD (1979) J Polym Sci Polym Symp 66:363
5. Van Wachem PB, Beugeling T, Feijen J, Bantjes A, Detmers JP, van Aken WG (1985) J Biomed Mater Res 6:403
6. Helmus MN, Gibbons DF, Jones RD (1984) J Biomed Mater Res 18:165
7. Ikada Y (1994) Biomaterials 13:725

8. Bures P, Huang Y, Oral E, Peppas NA (2001) J Control Rel 72:25
9. Kamath KR, Park K (1994) J Appl Biomat 5:163
10. Oehr C, Müller M, Elkn B, Vohrer U (1999) Surf Coat Technol 116:25
11. Lee JH, Park JW, Lee HB (1991) Biomaterials 12:443
12. Kiaei D, Hoffman AS, Horbett TA (1995) Radiat Phys Chem 46:191
13. Gupta B, Hilborn J, Hollenstein C, Plummer CJC, Xandropolous N, Houriet R (2000) J Appl Polym Sci 78:1083
14. Hayat U, Tinsley AM, Calder MR, Clarke DJ (1992) Biomaterials 13:801
15. Ward TL, Hinojasa O, Benesito RR (1982) Polym Photochem 2:109
16. Yeh Y, Iriyama Y, Matsuzawa Y, Hanson SR, Yasuda H (1988) J Biomed Mater Res 22:795
17. Ko T, Cooper SL (1993) J Appl Polym Sci 47:1601
18. Gupta B, Hilborn J, Bisson I, Frey P (2001) J Appl Polym Sci 81:2993
19. Sharma CP, Jayasree G, Najeeb PP (1987) J Biomat Appl 2:205
20. Hsiue G, Wang C (1993) J Polym Sci Chem 31:3327
21. Hsiue G, Lee SD, Wang CC, Chang PCT (1993) J Biomed Sci Polym 5:205
22. Sano S, Kato K, Ikada Y (1993) Biomaterials 20:529
23. Chen H, Belfort G (1999) J Appl Polym Sci 72:1699
24. Lee S, Hsiue G, Wang C (1994) J Appl Polym Sci 54:1279
25. Suzuki M, Kishida A, Iwata H, Ikada Y (1986) Macromolecules 19:1808
26. Ulbricht M, Belfort G (1996) J Membr Sci 111:193
27. Bae J, Seo E, Kang I (1999) Biomaterials 20:529
28. Kang I, Kwon OH, Lee YM, Sung YK (1996) Biomaterials 17:841
29. Kim EJ, Kang I, Jang MK, Park YB (1998) Biomaterials 19:239
30. Lee JH, Kopecek J, Andrade JD (1989) J Biomed Mater Res 23:351
31. Ko Yg, Kim YH, Park KD, Lee HJ, Lee WK, Park HD, Soo HK, Lee GS, Ahn DJ (2001) Biomaterials 22:2115
32. Fujimoto K, Inoue H, Ikada Y (1993) J Biomed Mater Res 27:1559
33. Desai NP, Hubbell JA (1991) J Biomed Mater Res 25:829
34. Gombotz WR, Guanghui W, Thomas AH, Hoffman AS (1991) J Biomed Mater Res 25:1547
35. Kang I, Kwon OH, Lee YM, Sung YK (1996) Biomaterials 17:841
36. Kato K, Yoshihito Y, Yamamoto M, Tomita N, Yamada S, Ikada Y (2000) J Adhesion Sci Technol 14:635
37. Uyama Y, Tadakaro H, Ikada Y (1991) Biomaterials 12:71
38. Richey T, Iwata H, Oowasaki H, Uchida E, Matsuda S, Ikada Y (2000) Biomaterials 21:1057
39. Eckert AW, Gröbe D, Rothe U (2000) Biomaterials 21:441
40. Uchida E, Uyama Y, Ikada Y (1990) J Appl Polym Sci 41:677
41. Jen MY, Ming CW, Ying GH, Chau HC, Sing KL (1998) J Membr Sci 138:19
42. Zhang F, Kang ET, Neoh KG, Tan KL (2001) Biomaterials 56:324
43. Chapiro A (1962) Radiation chemistry of polymeric systems. Wiley Interscience
44. Ratner BD, Hoffman AS (974) J Appl Polym Sci 18:3183
45. Hoffman AS, Achmer G, Harris C, Kraft WG (1972) Trans Am Soc Artif Intern Organs 18:10
46. Kabanov V, Kudryavtsev VN, Degtyareva TV, Zatikyan LL, Baskova IP, Starannikova LE (1997) Nucl Instr Methods Phys Res B131:291
47. Han DK, Jeong SY, Kim YH, Min BG, Cho HI (1991) J Biomed Mater Res 25:561
48. Brinkman E, Poot A, Van der Does L, Bantjes A (1990) Biomaterials 11:200
49. Kwon OH, Nho YC, Park KD, Kim YH (1999) J Appl Polym Sci 71:631
50. Gupta B, Büchi FN, Scherer GG, Chapiro A (1994) Polym Adv Technol 5:493
51. Gupta B, Anjum N (2000) J Appl Polym Sci 77:1331
52. Sun YH, Gombotz WR, Hoffman AS (1986) J Bioact Comp Polym 1:317
53. Gupta B, Tyagi PK, Ray AR, Singh H (1990) J Macromol Sci Chem 27:831
54. Singh DK, Ray AR (1994) J Appl Polym Sci 53:1115

55. Gupta B, Anjum N (2000) J Appl Polym Sci 77:1401
56. Park JS, Kim JH, Nho YC, Kwon OH (1998) J Appl Polym Sci 69:2213
57. Uchida K, Sakai K, Ito E, Kwon OH, Kikuchi A, Yamato M, Okano T (2000) Biomaterials 21:923
58. Kim YJ, Kang I, Huh MW, Yoon S (2000) Biomaterials 21:121
59. Gupta B, Plummer C, Bisson I, Frey P, Hilborn J (2002) Biomaterials 23:863
60. Gupta B, Anjum N (2002) J Appl Polym Sci 86:1118
61. Ratner BD (1995) Biosensors Bioelectron 10:797
62. Uyama Y, Tadokaro H, Ikada Y (1990) J Appl Polym Sci 39:489
63. Tomita N, Tamai S, Okajima E, Hirao Y, Ikeuchi K, Ikada Y (1994) J Appl Biomat 5:175
64. Ratner BD, Hoffman AS, Whiffen JD (1978) J Bioeng 2:313
65. Cohn D, Hoffman AS, Ratner BD (1987) J Appl Polym Sci 31:1
66. Lee HB, Shim HS, Andrade JD (1972) Am Chem Soc Polym Preprint 13:729
67. Alexander JW, Kaplan JZ, Altemeier WA (1967) Ann Surg 165:192
68. Singh H, Tyagi PK (1989) Die Angew Makromol Chemie 172:87
69. Tyagi PK, Gupta B, Singh H (1993) J Macromol Sci Pure Appl Chem A 30:303
70. Buckenska J (1996) J Appl Polym Sci 61:567
71. Park JS, Kim JH, Nho YC, Kwon OH (1998) J Appl Polym Sci 69:2213
72. Kang I, Kwon OH, Sung YK, Lee YM (1997) Biomaterials 18:1099
73. Yamata M, Konno C, Utsumi M, Kikuchi A, Okano T (2002) Biomaterials 23:561
74. Langer R, Vacanti JP (1993) Science 260:920
75. Temenoff JS, Mikos AG (2000) Biomaterials 21:431
76. Mirzadeh H, Katbab AA, Khorasani MT, Burford RP, Gorgin E, Golestani A (1995) Biomaterials 16:641
77. Kishida A, Iwata H, Tamada Y, Ikada Y (1991) Biomaterials 12:786
78. Kim B, Mooney DJ (1998) TIBTECH 16:224
79. Lee S, Hsiue G, Chang PC, Kao C (1996) Biomaterials 17:1599
80. Hilborn J, Frey P British Pat Appl 9,824,562
81. Tamada Y, Ikada Y (1994) J Biomed Mater Res 28:783
82. Gupta B, Frey P, Bisson I, Hilborn J (2000) Polym Preprints 4:1040
83. Burg KJL (2001) Intl Fibre J 16:38
84. Revagade N (2001) M Tech Thesis, IIT Delhi, India
85. Bisson I, Kosinki M, Rualt S, Gupta B, Hilborn J, Wurm F (2002) Biomaterials 23:3149
86. Kim H, Mayo U, Ko KT, Minoda M, Miyamoto T, Nakamure T (2001) Biomaterials 22:2489

Received: April 4th, 2002

The Effects of Radiation on the Structural and Mechanical Properties of Medical Polymers

Lisa A. Pruitt

Departments of Mechanical Engineering and Bioengineering, UC Berkeley, Berkeley, CA 94720 USA.
E-mail: lpruitt@newton.berkeley.edu

Medical polymers undergo changes in both structural and mechanical properties when sub-jected to ionizing radiation. Certain classes of polymers are susceptible to chain scission mechanisms while others are altered through crosslinking processes. These generalizations are complicated by the atmosphere in which the polymer is exposed to the ionizing treatment. Be-cause these devices must be sterilized prior to implantation their long-term structural evolu-tion owing to the effects of ionizing radiation coupled with environment must be understood. This work summarizes the wide-ranging effects of ionizing radiation on the mechanical and structural properties of medical polymers.

Keywords: Radiation, Medical polymers, Mechanical properties, Crosslinking, Scission, Aging, Oxidation

1	Sterilization of Implantable Devices	64
2	General Effects of Ionizing Radiation on Polymers	66
2.1	Interaction of Radiation with Polymers.	66
2.2	Molecular Changes	67
2.3	Free Radical Reactions	70
2.4	The Effect of Environment.	70
2.4.1	Absence of Oxygen	71
2.4.2	Presence of Oxygen.	72
2.4.3	Effects of Temperature	73
2.5	The Effect of Irradiation on Structure and Mechanical Properties . .	73
3	Specific Effects of Radiation on Medical Polymer Classes	75
3.1	Polyethylene.	75
3.2	Polypropylene.	76
3.3	Fluoropolymers	76
3.4	Polyacrylates and Polymethylmethacrylates	77
3.5	Nylons	78

Advances in Polymer Science, Vol. 162
© Springer-Verlag Berlin Heidelberg 2003

3.6   Polyurethanes   78
3.7   Polyesters and Biodegradable Polymers   79
3.8   Hydrogels   79

4   **Case Studies in Orthopedics**   80

4.1   Deterioration of Orthopedic Grade UHMWPE Due
       to Ionizing Radiation   80
4.2   Use of Ionizing Radiation and Low-Temperature Plasma Methods
       for Controlled Crosslinking of UHMWPE   85
4.3   The Effects of Ionizing Radiation on Acrylic Based Bone Cements . .   89

5   **Summary**   91

References   91

List of Abbreviations

DNA	Deoxyribonucleic acid
ESR	Electron Spin Resonance
EtO	Ethylene oxide
FTIR	Fourier transform infrared
FDA	Federal Drug Administration
LDPE	Low density polyethylene
HDPE	High density polyethylene
PET	Poly(ethylene terephthalate)
PGA	Poly(glycolic) acid (PGA)
PLA	Poly(lactic acid)
PMMA	Polymethylmethacrylate
PTFE	Polytetrafluoroethylene
PVDF	Poly(vinylidene fluoride)
PVF	Poly(vinyl fluoride)
UHMWPE	Ultra high molecular weight polyethylene

1
Sterilization of Implantable Devices

Sterilization of implantable devices is a critical concern to the medical industry. Such devices are either implanted for short duration in the body or they are implanted with the intent that they will achieve their function for several decades, and in some cases for the duration of the patient's life. This requires that the device must be sterile at the time of implantation to minimize risk of infection and that the sterilization method chosen must not degrade the properties of the biomedical material.

Infection is an important concern for the medical device industry. While the exact causes of infection remain unclear there appears to be an association of implanted biomedical devices with an enhanced likelihood for infection [1, 2]. The pathway for such infections is complicated and depends on numerous factors including the material used in the device, as well as the device location and expected implantation time [3]. For example a catheter associated with balloon angioplasty may be in place for minutes while a total joint replacement device may be expected to function in-vivo for two decades or more. Some materials appear to be more sensitive to bacterial contamination because of their composition, microstructure, degree of porosity, or surface chemistry of the material itself [4, 5].

Matthews et al. [1] define sterilization as the removal or destruction of all microorganisms from a contaminated material or device. However, the death of the microorganism is an exponential function of stress, and therefore it is typically defined by its capability to meet an end-point specification. In the medical device industry, the probability that the microorganism or bacterial contamination will survive a sterilization procedure defines its sterility. The FDA allows a device to be labeled as sterile if the number of non-sterile devices is less than one in a million (less than one device out of a lot of one million can show biological contamination). Sterilization methodology must provide pathways for spore death or must render the microorganism incapable of reproduction so that the device cannot support bacterial life in vivo.

The simplest sterilization techniques utilize elevated temperature. Steam is used for penetrable materials while dry heat is used for materials that are impenetrable to steam. Saturated steam sterilization utilizes temperatures in the range 120–135 °C for extended periods of time. Dry heat sterilization uses temperatures in the range 160–180 °C. This poses a unique problem to polymeric biomaterials, which are sensitive to the elevated temperatures. Sterilization can also be accomplished with chemical methods at lower temperatures. Low temperature steam can be coupled with formaldehyde and sterilization can also be accomplished by exposure to liquid glutaraldehyde. A limitation of these methods is that they can leave post-sterilization chemical residues on the polymeric devices. A chemical method, which has been quite successful in the medical industry, is ethylene oxide (EtO) gas (C_2H_2O) sterilization. This method uses an alkylation process to react with amino acids and protein groups for biocidal action. In the 1970s EtO was the most promising low temperature sterilization technology available. It was commonly used in medical devices and was adopted by numerous orthopedics and cardiovascular companies as the sterilization method of choice. EtO sterilization does have limitations however. It is a know carcinogen and must be handled with great care. Furthermore, devices sterilized with this gas must be out-gassed for several days in order to remove gaseous contaminants. For many medical companies, these limitations were sufficient to look for alternative sterilization methods for their polymeric devices. Several medical industries have now looked at low temperature gas plasma treatment as a viable sterilization technique [6]. Initially, however, the majority of the medical community and in particular the orthopedics industry switched to ionizing radiation for its sterilization efficiency and cost effectiveness.

Ionizing radiation is a method capable of delivering high effectiveness at relatively low cost. These treatments include gamma radiation, X-rays, or accelerated electrons. Cobalt 60 is used as the primary source for gamma radiation. Since its inception as a sterilization tool in the 1960s, gamma radiation sterilization is performed at a standard dose of 2.5 Mrad (25 kGy). Linear accelerators provide electron beams at energy levels that do not exceed 10 MeV but are limited in depth of penetration. The biocidal efficiency of ionizing radiation relies on free radical formation and its ability to diminish DNA replication in any bacterial spore or microorganism present in the medical device. A benefit of ionizing radiation is that it is carried out at ambient temperature, it can be through penetrating, and it requires no outgassing prior to implantation.

Ionizing radiation is being used at an increasing rate by the medical industry for the sterilization of devices comprising polymer components [7–9]. This has caused concern in the scientific community as polymers typically undergo some form of radiation-induced degradation. This can range from minor discoloration to severe oxidative embrittlement and deterioration of structural properties [10–12]. Understanding the reactions of radiation chemistry with various polymer classes used in biomedical devices is key to evaluating the longevity of devices sterilized by ionizing methods.

2
General Effects of Ionizing Radiation on Polymers

2.1
Interaction of Radiation with Polymers

In general, the exposure of polymers to ionizing radiation will alter their basic molecular structure and associated macroscopic properties. These molecular changes are brought about through a complex set of reactions upon exposure to radiation energy. Polymers are altered primarily through several basic schemes: electron absorption and subsequent cleavage that give rise to radical formation, radical combination resulting in the formation of crosslinks, or disproportionation to give scission, and gas evolution [13–17]. The deposition of high energy photons, such as gamma rays, in a polymer occurs via three physical mechanisms [18]: Compton scattering in which an electron is ejected while the photon is scattered, the photoelectric effect whereby a photon is absorbed and an orbital electron is ejected, and pair production in which a positron-electron pair is produced. In general, Compton scattering is the dominant activated in polymers.

There are several sources of ionizing radiation. These include gamma rays generated from a Cobalt-60 (1.17–1.33 MeV) or Cesium-137 source (0.66 MeV), and e-beams generated from electron accelerators (0.1–10 MeV) or Bremsstrahlung X-rays from accelerators (3–10 MeV) [19–23]. The standard unit of absorbed dose is the Gray (Gy) which is equivalent to the energy imparted by ionizing radiation to a mass of matter corresponding to one joule per kilogram. One Gray is also equivalent to 6.25×10^8 eV/kg. Another common unit of radiation dose, often encountered in the medical industry, is the rad, which is equivalent to the energy of absorption of 0.01 joule per kilogram (0.01 Gy).

2.2
Molecular Changes

Molecular changes brought about through ionizing radiation are commonly characterized with a G factor that quantifies the yield of an event [24, 25]. These events may include changes to molecular weight, gel content, or other molecular alterations. The G factor is defined as the event yield per 100 eV of energy and carries units of μmol/J. Most commonly, the G factors are described for crosslinking $G(X)$, scission $G(S)$, and gas evolution $G(g)$.

Molecular weight of a polymer, and its distribution, can be drastically altered by exposure to ionizing radiation. These high-energy treatments result in free radical generation, main chain scission and crosslinking, and subsequently alter the distribution of chain size in the bulk polymer. Much work has been done to develop the theoretical basis for molecular weight changes owing to high-energy radiation exposure [26–29]. For a linear polymer enduring chain scission, the basic form of the equation describing the change in molecular weight distribution is given as [29]

$$\frac{\partial \phi(p,y)}{\partial y} = -p\phi(p,y) + 2p \int_p^\infty \frac{\phi(l,y)}{l} dl$$

where

$$y = \int_0^t r \, dt$$

r is the probability that a polymer chain undergoes scission in unit time, p is the degree of polymerization, and $\phi(p,y)$ is the weight fraction of polymer molecules having p structural units. The term on the left describes the decrease in molecules having p structural units due to main chain scission. The term on the right describes the increase of the molecules having p structural units due to scissions of those molecules having 1 units. The solution to the above equation takes the form [29]

$$\phi(p,y) = \left\{ \phi(p,0) + py \int_p^\infty \frac{(2+yl-yp)}{l} \cdot \phi(l,0) dl \right\} \exp(-py)$$

where $\phi(p,0)$ is the initial weight fraction. The G factor for scission, $G(S)$ takes the form [29]:

$$G(S) = 100 \frac{N_A}{R_d} y$$

Where R_d is the radiation dose and N_A is Avagadro's number. The number average degree of polymerization, P_n, in the absence of any crosslinking is given as [29]:

$$P_n^{-1} = u_n^{-1} + y$$

where u_n is the number average degree of polymerization prior to irradiation. If $M_{n,i}$ and M_n are the number average molecular weight prior to and subsequent to irradiation, respectively, then:

$$u_y = \frac{(M_{n,i} - M_n)}{M_n}$$

This indicates that the alterations to the *number* average molecular weight depend only on $M_{n,i}$ and the scissions per structural unit rather than on the initial molecular weight distribution. However, changes to the *weight* average molecular weight resulting from radiation-induced scission do depend on the initial molecular weight distribution. For the case of an initial uniform distribution, the relative change in weight average molecular weight takes the form [29]:

$$\frac{M_w}{M_{w,i}} = \frac{2(u_n - 1 + e^{yu_n})}{(u_n y)^2}$$

For the case of an initial random distribution, the relative change is given as:

$$\frac{M_w}{M_{w,i}} = \frac{1}{(1 + u_n y)}$$

When only degradation occurs within the polymer, the evolution of molecular weight distribution will approach that of a random case, such that $M_w/M_n = 2$.

Under conditions of simultaneous crosslinking and scission, the G factor takes the form:

$$G(X) = 100 \frac{N_A}{R_d} x$$

Where x is the number of crosslinks per structural unit, or crosslink density. If scission and crosslinking occur at the same time it is typically assumed that they are independent events. That is, the evolution in molecular weight can be determined by summing the effect of scission and then subsequently the effect crosslinking. For such cases, the *number* average molecular weight changes is described as [29]:

$$\frac{M_n}{M_{n,i}} = \frac{1}{1 + u_n(y - x)}$$

As before, the number average molecular weight is not a function of the initial molecular weight distribution. The weight average molecular weight, however, does depend on the initial distribution. For an initial random distribution, the relative change in *weight* average molecular weight is given as:

$$\frac{M_w}{M_{w,i}} = \frac{1}{1 + u_n(y - 4x)}$$

and for a uniform initial distribution:

$$\frac{M_w}{M_{w,i}} = \frac{2(e^{-u_n y} + u_n y - 1)}{\left[u_n y^2 - 4 u_n x(e^{-u_n y} + u_n y - 1) \right]}$$

For the case of crosslinking without scission effects, the polymer will increase the molecular weight until its value becomes infinitely large. As this occurs, gelation occurs in the portion of the polymer that is insoluble in any solvent, increases with radiation dose. The point at which the very same gel is formed is termed the gel point and the concomitant crosslink density at this point, x_g, is defined as [29]:

$$x_g = \frac{1}{2P_{w,i}}$$

The number of crosslinks at the gel point is equal to half the ratio of the total number of structural units to the weight average degree of polymerization prior to crosslinking, $P_{w,i}$.

2.3
Free Radical Reactions

For polymeric materials the primary mechanism of damage occurs via free radical formation. Parkinson [18] has outlined the basic steps for the generation of reactive free radicals resulting from ionizing treatment. The damage is initiated with the ejection of a high-energy electron. The reaction is depicted below:

$R \Rightarrow R^+ + e-\}$ generation of free electron

$e- + R \rightarrow R^+ + 2e-\}$ ionization

$R^+ + e- \rightarrow R^*\}$ highly excited electron states

These highly excited electron states may undergo radiationless or radioactive decay. They may also decay through a chemical pathway via heterolytic bond cleavage leading to the generation of ions or via homolytic bond cleavage resulting in the generation of free radicals:

$RX^* \rightarrow R^\bullet + X^\bullet\}$ free radical formation

In general, for polymers, it is the free radical generation that dominates. This leads to free radical reactions that depend on the chemistry of the polymer and may be coupled with the environment. Such reactions can result in main chain scission, recombination, or disproportionation as exampled below in polyethylene:

$-(CH_2 - CH_2)- \rightarrow -(HC_2^\bullet + {}^\bullet CH_2)-\}$ main chain scission

$-({}^\bullet CH_2 + {}^\bullet CH_2)- \rightarrow -(CH_2 - CH_2)-\}$ recombination

$-(CH_2 - {}^\bullet CH_2 + {}^\bullet CH_2)- \rightarrow -(CH = CH_2 + CH_3)-\}$ disproportionation

If the side group of the polymer is involved in scission, small radical fragments are generated, such as a hydrogen radical. The danger in this reaction is that this small radical is highly reactive and mobile, and it may be able to diffuse great distances triggering numerous reactions.

2.4
The Effect of Environment

Ionizing treatment can result in chain scission or crosslinking depending on the environment in which the radiation is performed, and the availability of reactive gases or radicals within the polymer itself. Further, susceptibility to one degradation scheme or the other is dictated by the intrinsic backbone chemistry of the polymer. Many polymers undergo both scission and crosslinking, but often one mechanism predominates.

2.4.1
Absence of Oxygen

The nature of the polymer structure determines its response and physical change upon exposure to ionizing radiation. In general, as shown below, the polymers of Type (A) will undergo a predominant scission mechanism while polymers of Type (B) will be susceptible to crosslinking:
- Type A: $(CH_2-CR' R'')$ favors scission
- Type B: (CH_2-CHR) favors crosslinking

As an example, poly(methylmethacrylate) (PMMA) is representative of a Type A polymer with side groups, $R'=CH_3$ and $R''=COOCH_3$ and it predominately degrades via chain scission. On the other hand, poly(methyl acrylate) is a Type B polymer with side group $R=COOCH_3$ and it readily crosslinks. In general, polymers with high concentrations of quaternary carbon atoms along the chain undergo scission while polymers lacking this feature predominantly crosslink. The presence of unsaturation in the polymer chain aids this effect and enhances the yield of crosslinking (natural rubber is an example). Also, the presence of aromatic groups lower the yield of any reaction due to radiation (polystyrene is one of the most radiation resistant polymers). Table 1 provides a summary of the general polymer schemes that are susceptible to either crosslinking or scission.

Table 1. Summary of the general polymer schemes which are susceptible to either crosslinking or scission when subject to ionizing radiation in the absence of oxygen

Polymers prone to scission	Polymers prone to crosslinking
Poly(methyl methacrylate)	Polyethylene
Polyisobutylene	Polypropylene
Poly α-methylstyrene	Polystyrene
Poly(vinylidene chloride)	Polyacrylates
Poly(vinyl fluoride)	Polyamides
Polyacrylonitrile	Poly(vinyl chloride)
Polytetrafluorethylene	Polyesters
Cellulose and derivatives	Unsaturated elastomers
DNA	Natural rubber

2.4.2
Presence of Oxygen

The ambient environment drastically affects the response of a polymer to radiation exposure. When ionizing radiation is performed in air there can be a severe propensity for structural degradation of the polymer. Most polymers, even those that crosslink in an inert environment, will have a tendency to undergo chain scission when oxygen is available during ionization [30, 31]. In such cases, the general classification described in Table 1 is no longer valid. Also damage often occurs at much lower radiation doses, and degradation becomes a time dependent process.

The time dependent mechanism of free radical reactions poses serious concern for radiation degradation of polymers, particularly in the presence of oxygen due to its high diffusional mobility and reactivity. Parkinson [18] has outlined the free radical reactions in radiation-induced polymer oxidation as follows:

$$R - \gamma \rightarrow R^{\bullet} \} \text{ initiation}$$

$$R^{\bullet} - O_2 \rightarrow RO_2^{\bullet} \} \text{ propagation}$$

$$RO_2^{\bullet} - RH \rightarrow RO_2H + R^{\bullet} \}$$

$$RO_2H - RH \rightarrow RO^{\bullet} + {}^{\bullet}OH \} \text{ chain branching}$$

$$RO^{\bullet} - RH \rightarrow ROH + R^{\bullet} \}$$

$${}^{\bullet}OH - RH \rightarrow H_2O + R$$

$$RO_2H, RO_2^{\bullet}, R^{\bullet} \rightarrow \text{ scission and crosslinking}$$

$$2RO_2^{\bullet} \rightarrow RO_2R + O_2 \} \text{ termination}$$

The high diffusional mobility of oxygen provides a greater penetration into polymer and results in the generation of reactive free radicals that can be long-lived. Oxygen traps free radicals and drives reactions that are oxidative in nature. These chemical reactions result in the generation of oxidative products created within the polymer and include the formation of carboxylic acids, ketones, esters, alcohol derivatives, and peroxide based species [32, 33]. Free-radical reactions result in gaseous products, and in the presence of oxygen these can include carbon dioxide, carbon monoxide, and water vapor. When peroxides degrade, chain branching is favored and reactions result in further radical generation. This provides a pathway for the generation of numerous free radicals within the polymer. Additionally, the peroxide decomposition is temperature dependent. This results in a polymer that has both a time and temperature dependent degradation mechanism. Moreover, the depth of oxidation within the polymer depends on the oxygen-permeation rates and consumption rates per absorbed dose as well as the dose rate itself. Thus the polymer has far less stability than its counter part radiated in an inert environment.

Such post-irradiation degradation schemes can render a polymer both oxidized and embrittled [34]. This has serious consequences for the medical device industry that uses gamma radiation for sterilization and relies on stability of shelf storage prior to implantation. In such instances the details of the sterilization environment and subsequent shelf storage become paramount to long-term integrity of the polymer.

2.4.3
Effects of Temperature

The temperature during irradiation can have a profound effect on the radiation chemistry. Below the glass transition temperature, a large number of stable radicals are generated and trapped in the glassy state [35]. Correspondingly, this reduces the net crosslinking due to immobility in the glassy state. Above the glass transition temperature, the tendency toward crosslinking or scission is enhanced depending on the polymer chemistry. For example, in polyethylene if irradiation is done at liquid nitrogen temperatures allyl double bonds are prevented from forming. On the other hand, if polyethylene is irradiated in the melt the yield of crosslinks is greatly enhanced. Polystyrene shifts its tendency from crosslinking to scission as the temperature is brought above the glass transition temperature. In some fluoropolymers, the opposite trend is observed. Crosslinking is favored at temperatures above the glass transition. The effect of temperature can be further accelerated in the presence of mechanical stresses on the material [36]. This is likely the result of weakened chemical bonds under stress and the plasticizing effect of the polymer via radiochemical products.

2.5
The Effect of Irradiation on Structure and Mechanical Properties

During irradiation, semicrystalline polymers undergo structural changes resulting in imperfections in the crystalline phase. In general, as higher degrees of radiation are sustained the crystalline structure evolves to a degraded structure resulting in an accompanying decrease in the crystalline melting point, Tm. Poly(ethylene terephthalate) exhibits a 25 K depression in melting point when radiated to a dose of 2000 Mrad [37]. In some instances, the polymer will increase its net crystallinity and density with increasing radiation. An example of this is ultra high molecular weight polyethylene (UHMWPE). Because UHMWPE has a very high molecular weight (~4–6 million g/mol) and a relatively low crystallinity (~50%), the result of ionizing radiation is a scission-based process resulting in smaller chains. The smaller chains pack more efficiently and result in an increased density and crystallinity upon irradiation [38]. Similar effects are observed in PTFE and poly(vinylidene fluoride) PVDF. The effect of increasing crystallinity and density can alter basic mechanical properties such as yield stress and modulus.

In general, for flexible polymers that experience crosslinking as a result of ionization, the elastic modulus tends to increase while the strain to failure decreases. With increasing dose, the polymer becomes stiffer, harder and at a sufficiently high dose it becomes brittle. Figure 1 shows the effect of radiation dose on a polychloroprene polymer. For flexible polymers susceptible to scission, the mechanical properties such as tensile strength, modulus, and elongation decrease with dose. For these polymers the mechanical properties are continuously diminished and may ultimately evolve into viscous liquids. Flexible polymers and elastomers are commonly susceptible to radiation-induced damage while glassy polymers and crosslinked resins tend to be more resilient. Glassy polymers that undergo

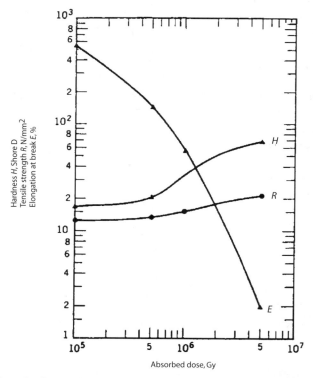

Fig. 1. The effect of radiation dose on the mechanical properties of a polychloroprene polymer. E denotes elongation at break, H represents Shore D hardness, and R represents the tensile strength [18]

crosslinking are resistant to deterioration and experience little change in their mechanical properties. Modulus is often retained in glassy polymers that are susceptible to scission but tensile strength is lost with increasing radiation dose. However, at low crosslinking levels brought about through irradiation, polystyrene can be altered from a glassy polymer to a rubbery material as the polymer behaves like an elastic network above its glass transition temperature [39]. Crosslinked, thermosetting plastics are highly resistant to radiation damage because of their high concentration of polar groups and intermolecular dipole-dipole interactions.

3
Specific Effects of Radiation on Medical Polymer Classes

3.1
Polyethylene

Polyethylene has numerous applications in the medical industry ranging from catheters to bearing surfaces for joint replacements [40, 41]. As one of the most viable commercial polymers, polyethylene has been the subject of extensive radiation studies. In fact, there is a plethora of research devoted to the crosslinking of polyethylene for commercial applications ranging from paraffins to UHMWPE [38, 42–47]. As previously described, when polyethylene is exposed to an ionizing radiation environment in the absence of oxygen it will favor crosslinking. The crosslinks generated are located primarily in the non-crystalline phase and along the lamellar-amorphous interphase [42]. When polyethylene is irradiated in the melt a greater density of crosslinks can be created at the expense of the crystalline phase. In general with increasing crosslink density there is a concomitant reduction in crystallinity. Further, as a result of the backbone structure of the polyethylene chain, there will be several unique responses to ionizing radiation including the generation of hydrogen gas, the unsaturation of transvinylene bonds, the degradation of vinyl end groups, and capability for high crosslink efficiency. Trans vinylene unsaturation does not depend on the molecular weight but is linearly dependent on the radiation dose making it a successful dosimeter [48].

In general, ionizing radiation via e-beam or gamma rays that results in crosslinking is considered beneficial to the mechanical properties of polyethylene. Moderate crosslinking in HDPE results in an increase in both yield strength and tensile modulus [38]. For UHMWPE, radiation in the melt results in a high yield of crosslinking with little scission. With increasing crosslink density there is a corresponding evolution of crystallite morphology from lamellar to micellar [49]. Crosslinking of UHMWPE is of great interest to the orthopedics community. Recent studies on the wear behavior of highly crosslinked UHMWPE used as the bearing surface for total joint replacements suggests that crosslinking could drastically improve the wear resistance of UHMWPE [50–54].

On the other hand, if precautions are not taken to prevent the presence of molecular oxygen at the time of irradiation, polyethylene will undergo a dominant scission mechanism. This has lead to numerous embrittlement problems associated with oxidation and chain scission [9–12]. This radiation chemistry plagued the orthopedics community for years when many medical device companies utilized gamma radiation in air as the primary sterilization method [55].

A case study describing the effects of ionizing radiation on UHMWPE and the implications for the medical community is provided below.

3.2
Polypropylene

Polypropylene, in its isotactic semicrystalline form, is commonly used in non-resorbable sutures. This thermoplastic polymer undergoes both crosslinking and scission mechanisms when irradiated. This polymer suffers from post-radiation deterioration of mechanical properties regardless of whether oxygen is present or absent at the time of radiation [56]. However, oxidative degradation dominates this post radiation evolution in physical properties and results in a reduction of modulus, tensile strength, and strain to failure [57]. The loss of strain to failure follows first order kinetics and results in a reduction from 900% elongation to failure to values approaching 20% at a radiation dose of 160 Mrad [58, 59].

3.3
Fluoropolymers

Fluoropolymers are fully or partially fluorinated polyolefins that provide chemical inertness, thermal resistance, mechanical durability, and lubricity. These unique properties are due to the fluorine sheath that protects the carbon backbone from chemical attack. Most fluorocarbon polymers are semicrystalline due to the spatial packing efficiency of the linear or helical polymer chains. Fluoropolymers are used in a wide variety of medical applications ranging from vascular grafts to microfluidic devices [60]. While there are a wide range of fluoropolymers available, the use of these materials for biomedical applications is predominantly based on polytetrafluoroethylene in expanded form. The success of this biomaterial has resulted from its micro porous structure, which allows bio-integration for fixation and provides structural integrity. The most successful medical applications of e-PTFE are found in cardiovascular surgery where e-PTFE vascular grafts are used for vessel reconstruction. In orthopedic surgery, e-PTFE has been used as a replacement for the anterior cruciate ligaments in the knee. As with all medical devices it is likely that this polymer can be subjected to ionizing radiation as a standard sterilization scheme.

Fluoropolymers, given their chemical and structural inertness, are surprisingly susceptible to post-radiation degradation [61]. Moreover, the resulting response to ionizing radiation depends on the backbone chemistry of the fluoropolymer. As a general rule, there is a greater likelihood for crosslinking with increased hydrogen on the polymer backbone. So PVF will favor crosslinking while PTFE will be susceptible to degradation. The latter is very susceptible to scission processes, and in fact ionizing radiation can be used to tailor the molecular weight of this polymer commercially [62]. Similar to polyethylene, PTFE is plagued with peroxy-based free radical generation when exposed to ionizing radiation, and such radicals are long-lived in the polymer. Hedvig [63] has investigated the tensile properties of gamma radiated PTFE and has found a 90% reduction in strain to failure and nearly a 40% reduction in tensile strength. Like polyethylene, the crystallinity of PTFE is found to increase with increasing radiation dose up to 100 Mrad due to scission mechanisms. Beyond this dose point the crystallinity is reduced due to an increase in free volume of the polymer [63]. Deterioration of structural properties of PTFE can be minimized if radiation is performed in a vacuum environment. In this case crosslinking mechanisms can be activated and increases in tensile strength can be achieved.

3.4
Polyacrylates and Polymethylmethacrylates

Polyacrylates and polymethylmethacrylates are commonly used as medical adhesives, and can be found commonly in the form of bone cements and dental resins [40]. While these polymers appear to be quite similar in their backbone chemistry, as mentioned above, there are subtle differences that render very different reactions to ionizing radiation. Poly(alkyl acrylates) undergo crosslinking when exposed to radiation, while poly (alkyl methacrylates) degrade very rapidly through scission schemes. In fact the latter degrades so efficiently that this polymer is used as a positive electron beam resist [64]. Similarly polyacrylates are quite stable when radiated by ionizing methods; however polymethylmethacrylate is susceptible to scission and is susceptible to severe loss in strength as a consequence [18]. This degradation and the accompanying loss of mechanical properties are illustrated in Fig. 2. Dickenson [65] has shown that the number of main chain scissions is proportional to radiation dose in PMMA. A case study describing the effects of ionizing sterilization on the structure and mechanical properties of acrylic based bone cements is provided below.

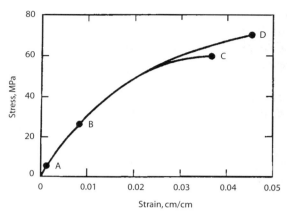

Fig. 2. Stress-strain behavior for PMMA exposed to four doses of gamma radiation in an inert environment at 25 °C: (A) 40 Mrad, (B) 20 Mrad, (C) 4 Mrad, (D) unirradiated [18]

3.5
Nylons

Nylons are commonly used as a suture material in the medical industry and in some instances they are used in polymeric balloons for balloon angioplasty [40]. On the whole, nylon is characterized as a crosslinkable polymer but this class of polyimides is susceptible to both scission and crosslinking. This behavior has been linked to the number of methylene groups or hydrogen atoms on the polymer backbone [66]. These polymers, like polyethylene, are drastically affected by the environment in which irradiation is performed. In the absence of oxygen there is little change in tensile strength with increasing dose; however, in the presence of air there is substantial drop in both strength and elongation to break. Nylons can be stabilized with the addition of an aromatic group – the aliphatic polyimides are quite resilient to radiation damage.

3.6
Polyurethanes

Thermoplastic polyurethanes have broad applicability in the medical plastics in-dustry and most notably in catheters [67]. Thermoplastic polyurethanes are block copolymers that comprise amorphous and crystalline blocks. The former dictates the elasticity while the latter determines the stiffness and strength. The amorphous blocks are often composed of ethers or esters. The ethers are utilized where low temperature flexibility, microbial resistance, and hydrolytic stability are needed, while esters are used where tear strength, abrasion resistance, and toughness are required [67]. In general, crosslinking via radiation is used to improve the strength, abrasion resistance, and modulus of these polymers. This enables the polymer to be injection molded into complex shapes and then cured via ionizing radiation.

3.7
Polyesters and Biodegradable Polymers

Polyesters are used in numerous biomedical applications. Poly(ethylene terephthalate) also known as PET is used predominantly in woven fabric form for cardiovascular prostheses. They are used for arterial patches and large diameter vascular prostheses [40]. Aliphatic polyesters such as polycaprolactone are biodegradable polymers with applications ranging from resorbable sutures to controlled drug delivery devices. Poly(glycolic) acid (PGA) and poly(lactic acid) (PLA) are the most widely used bioerodable polymers. PGA is used in resorbable sutures and offers enhanced strength due to its relatively high crystallinity. The amorphous form of PLA is used for drug delivery while the semicrystalline form is used in degradable sutures or for resorbable load bearing devices such as orthopedic screws. The mechanism of biodegradation is broadly categorized into three mechanisms: cleavage of crosslinks between water soluble chains, cleavage of side chains leading to the formation of polar groups, or main chain cleavage. Because scission is the predominant response of these polymers to radiation, great care must be made when sterilizing these polymers. Ionizing radiation can alter the rate of degradation and can lead to premature failure of these devices. For this reason many resorbable polymers are sterilized with EtO or plasma methods.

3.8
Hydrogels

Hydrogels are crosslinked water-swollen polymeric structures that are commonly used in drug delivery and tissue engineering applications [40]. Hydrogels are made by swelling crosslinked polymers in biological fluids or water. The crosslinked structure is commonly prepared with ionizing methods such as gamma rays, X-rays, e-beam, or ultraviolet light. Because ionizing methods are used for the crosslinking mechanism, sterilization is typically done with non-ionizing methods such as EtO or plasma so that the crosslinking dose and structural integrity of the hydrogels can be controlled.

4
Case Studies in Orthopedics

The use of polymers in orthopedics requires an implant material that tolerates large fluctuating stresses due to contact loads, offers excellent tribological performance, and provides long-term structural stability. For this reason the only successful polymer used at the articulating surface is ultra high molecular weight polyethylene (UHMWPE). Because of its exceptionally high molecular weight and tie molecule density this polymer offers exceptional fatigue, wear, and abrasion resistance. Further, in its bulk form it is highly biocompatible. It is susceptible to damage, however, under long-term cyclic loads in vivo. This has been found to be exacerbated by the effects of ionizing radiation when used as a sterilization technique. Similar effects have been found in acrylic based bone cements, which are used to integrate the orthopedic device into the surrounding bone tissue. Studies addressing the effect of ionizing radiation in orthopedic polymers have greatly contributed to the understanding of the structural evolution resulting from exposure to radiation sources associated with sterilization and processing. These studies are reviewed below.

4.1
Deterioration of Orthopedic Grade UHMWPE Due to Ionizing Radiation

Sterilization via ionizing radiation has been the primary culprit in the long-term degradation of UHMWPE used for total joint replacements. As recently as 1995, UHMWPE was customarily sterilized with gamma radiation in the presence of air using a nominal dose of 25 kGy. In 1998, most United States Orthopedic Companies had transitioned their sterilization techniques to one of a few methods including gamma radiation in an inert environment or a non-ionizing sterilization method using ethylene oxide or gas plasma. This change in sterilization practice was motivated by the plethora of scientific literature indicating that gamma sterilization in air promoted oxidative chain scission and long-term degradation of desirable physical, chemical, and mechanical properties of UHMWPE [55, 68–76]. Major orthopedic companies have discontinued sterilization of UHMWPE components using gamma radiation in the presence of air; however, the long-term deterioration of this polymer remains a clinically relevant problem. The reason for this continued challenge is that an estimated four million U.S. patients have been implanted with an UHMWPE component that was sterilized in air during the period 1980–1995, when gamma irradiation in air was the regular sterilization practice.

As discussed above, free radicals generated by ionizing radiation within UHMWPE are vulnerable to oxidation during shelf aging and after implantation. Premnath et al. [55] suggested that shelf aging in air represents a more extreme oxidative challenge for irradiated UHMWPE than in vivo oxidation, because the concentration of dissolved oxygen in air has approximately an order of magnitude greater concentration of oxygen than that dissolved in the surrounding body fluids under physiological conditions. Radiation sterilization of UHMWPE in air, followed by long-term shelf aging, is now understood to result in degradative changes to the physical, chemical, and mechanical properties of the polymer [77–79]. For this reason shelf aging has become the focus of numerous recent studies in an attempt to understand and replicate the oxidation of UHMWPE. Post-irradiation aging has been simulated using a combination of thermal conditioning and elevated oxygen partial pressures. Accelerated aging protocols have been employed not only to differentiate the effects of gamma sterilization in air, but also to evaluate the oxidation resistance of UHMWPE sterilized by alternative methods.

There are two currently accepted testing strategies that are considered as viable accelerated aging protocols for UHMWPE. The first method developed by Sun et al. preconditioned specimens in a standard air furnace at 80 °C for up to 23 days, albeit at a controlled, slow initial heating rate of approximately 0.6 °C/min [80]. This study utilized a control group of UHMWPE components that had been gamma radiation sterilized in air prior to accelerated oxidation or shelf aging for up to 10 years. Sun et al. found that, after 23 days of preconditioning, the maximum oxidation index in the specimens corresponded to that of 7–9 years of shelf aging. Although Sun and colleagues matched the peak oxidation index between the preconditioned and shelf aged components, a subsequent study showed that the distribution of oxidation through the thickness differed between the two groups [81]. The shelf-aged components showed a subsurface oxidation peak, while the artificially aged components had a maximum oxidation at the surface. The second method devised by Sanford and Saum in 1995 [82] employed a rapid technique for accelerating oxidative degradation in UHMWPE. Specimens were aged in pure oxygen under five atmospheres of pressure at 70 °C for up to seven days in order to achieve a subsurface oxidation peak, similar to that observed in retrieved components. Comparing the bulk average oxidation index from preconditioned components with that from shelf-aged components validated their method. It was found that one week of preconditioning at 70 °C and five atmospheres of pure oxygen resulted in oxidation equivalent to five to ten years of ambient shelf aging. It should be noted that accelerated oxidation methods for UHMWPE are not without their limitations. For example, in a morphology study of UHMWPE that was gamma-radiation sterilized in air and then preconditioned using Sun's method, Crane et al. [83] found the crystalline morphology in the preconditioned specimens was notably different than in shelf-aged components. Further research is needed on the mechanical behavior and morphology of UHMWPE to elucidate the differences between thermal conditioning and long-term shelf aging.

The level of oxidation in aged UHMWPE components that have aged on the shelf, in-vivo, or through accelerated means can be quantified using spectroscopic techniques. Kurtz et al. performed numerous validation studies to ensure reproduc-

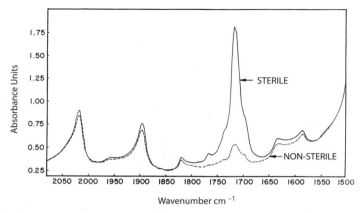

Fig. 3. Two infrared spectra for UHMWPE, one unirradiated and the other sterilized by gamma radiation and highly oxidized. Note that in the highly oxidized sample there is absorbance in the wave number range of 1680–1800 cm^{-1} which is the carbonyl peak containing ketones, esters, aldehydes, and acids

ibility of spectroscopic characterization [76]. In the orthopedics community infrared spectroscopy has been the chosen analytical tool for the quantification of oxidation. Figure 3 shows two infrared spectra from UHMWPE, one unirradiated and the other highly oxidized. Note that in the highly oxidized sample there is absorbance in the wave number range of 1680–1800 cm^{-1} which is known to be due to carbonyl containing chemical species, such as ketones, esters, aldehydes, and acids [84]. Calculating the area under the vibrational peak can provide quantification of the carbonyl vibration. This value is normalized to obtain a dimensionless number.

Primary attention is given to oxidation embrittlement which results in the decreased wear and fatigue resistance of the polymer. Researchers have shown gamma radiation in air to be the primary culprit in degrading the mechanical and structural integrity of UHMWPE [85]. There has been a strong correlation between delamination wear and surface damage with oxidation levels in the polymer. Recent work by Baker et al. [86] has shown that ionizing radiation results in loss of fatigue crack propagation resistance in UHMWPE. Further deterioration was observed when the polymers were subjected to accelerated aging. Figure 4 shows the effect of sterilization method and the effects of accelerated aging on the fatigue crack propagation resistance of UHMWPE. Scanning electron microscopy was used to provide an understanding of fatigue fracture mechanisms. Figure 5 shows the accompanying scanning electron micrographs depicting a change from the (a) ductile tearing observed in the non-sterilized (control) UHMWPE to the (b) cellular microstructure observed in the highly embrittled UHMWPE subjected to gamma radiation and accelerated aging in the presence of oxygen. It was found in

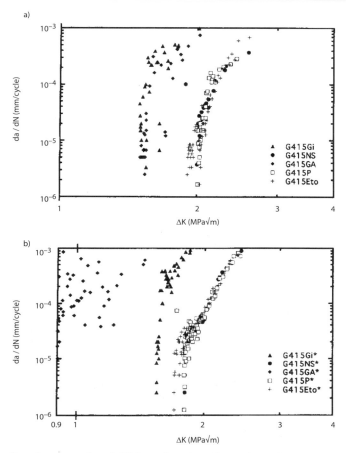

Fig. 4. Plot of crack propagation, da/dN, as a function of stress intensity range or crack driving force, ΔK. The plot shows the effect of sterilization method and the effects of accelerated aging on the fatigue crack propagation resistance of UHMWPE [86]

this work that accelerated aging caused a decrease in fatigue resistance regardless of sterilization method. This work demonstrated that loss of fatigue resistance was most severe for gamma radiation in air coupled with accelerated aging conditions. Another study by Goldman and Pruitt [6] has shown that non-ionizing sterilization techniques such as EtO and low-temperature gas plasma retain the highest ductility in the polymer and the associated resistance to growth of fatigue cracks.

At this time there is no clear scientific agreement as to which sterilization methods will provide the most beneficial long-term performance of the UHMWPE implant. All of the sterilization methods currently employed by the orthopedic community fulfill their intended purpose, namely the elimination of bacterial spores that could lead to infection and premature revision. Assessment of the specific

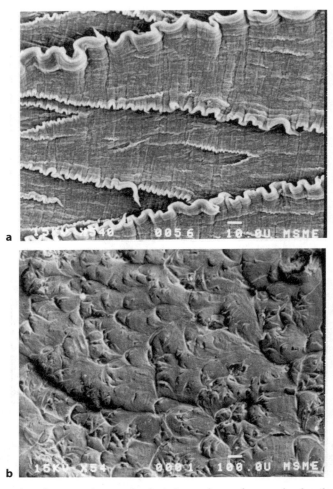

Fig. 5a,b. Scanning electron micrographs depicting a change from: **a** the ductile tearing observed in the non-sterilized (control) UHMWPE to; **b** cellular microstructure observed in the highly embrittled UHMWPE subjected to gamma radiation and accelerated aging in the presence of oxygen

sterilization methods has been complicated by the potential benefit of combining the manufacturing and sterilization processes. For example, radiation doses in excess of 25 kGy are known to provide crosslinking in UHMWPE if performed in an inert environment. Thus it becomes difficult to distinguish sterilization and processing when ionizing radiation is employed as the crosslinking medium for UHMWPE (discussed below).

4.2
Use of Ionizing Radiation and Low-Temperature Plasma Methods for Controlled Crosslinking of UHMWPE

Recent studies in the orthopedic community have suggested that a high degree of crosslinking can benefit the wear resistance of UHMWPE [87–89]. These initial studies have led to a more thorough investigation into the effects of crosslinking on the wear behavior of UHMWPE [90–92]. Based on the findings of these early studies the efforts of the orthopedics community have shifted towards the manufacture of highly crosslinked UHMWPE network structures with a reduced concentration of residual free radicals. These crosslinked polyethylenes recently received FDA approval in the United States for use in total hip and knee arthroplasty, and the next few decades will determine their long-term success in-vivo.

While crosslinked UHMWPE appears to be an innovative trend in orthopedics, there are several early examples of investigations on highly crosslinked UHMWPE for bearing surface applications. One of the early attempts on the use of crosslinked UHMWPE in total hip arthroplasty was by Oonishi et al. who reported on improved wear resistance of UHMWPE crosslinked with high doses (100 Mrad) of gamma radiation [89, 93]. Their clinical studies used four groups of patients that are described in Table 2 [76]. The radiographic wear measurements showed average rates of femoral head penetration of 0.076 mm/year for the Co-Cr alloy femoral head/irradiated UHMWPE pair and 0.072 mm/year for the alumina/irradiated UHMWPE. In comparison, with unirradiated UHMWPE liners the penetration rates were 0.247 and 0.098 mm/year for the Co-Cr and alumina head groups respectively.

The current research in orthopedics is centered around the most favorable processing conditions and the optimization of crosslink density for wear resistance

Table 2. Combined clinical wear data of the work of Oonishi et al. [89, 93] on γ-irradiated UHMWPE

Group	Femoral component (head size)	Acetabular component	Number of patients[a]	Average follow-up (months)	Average wear rate (mm/year)
I	COP alloy[b] (28 mm)	γ-Irradiated UHMWPE	19	12	0.076
II	Alumina (28 mm)	γ-Irradiated UHMWPE	9	6	0.072
III	Co-Cr[c] (28 mm)	Conventional UHMWPE	15	8	0.247
IV	Alumina (28 mm)	Conventional UHMWPE	71	6	0.098

[a]Extracted from the graphs of references [76, 89, 93]; [b]Stainless steel with 20% cobalt; [c]T-28 by Zimmer

of the bearing surface. Jasty et al. [94] described the wear behavior of a new form of UHMWPE (IMS-200). The acronym stands for *i*rradiated in the *m*olten *s*tate to *200* kGy. In their study, the polymer was heated to 140 °C and crosslinked in its molten state using electron beam irradiation to a total absorbed dose level of 200 kGy. In hip joint simulator experiments carried out to 5 million cycles, there was no detectable weight loss in the crosslinked components. Moreover, Jasty and coworkers stated that the initial machining marks were still present on the articulating surfaces of the liners after being subjected to 5 million gait cycles on the hip joint simulator.

Muratoglu et al. [95] reported on the effect of crosslink density (radiation dose) on the wear behavior and mechanical properties of UHMWPE. In their study, UHMWPE was crosslinked in air using electron beam irradiation to varying dose levels and was subsequently melt-annealed in the molten state at 150 °C for 2 h. The process was named CISM and the acronym stands for *c*old *i*rradiation with *s*ubsequent *m*elt-annealing. Melting was used to increase the chain mobility and enhance the recombination reactions between the residual free radicals and hence reduce their concentration. The ESR experiments showed no detectable residual free radicals. Some properties were found to be independent of radiation dose level. The melting point of the polymer remained unchanged as a function of absorbed radiation dose. The crystallinity decreased from 54% to 48% upon the initial CISM treatment but remained unchanged with increasing radiation dose. The yield strength behaved in a similar way and decreased by about 14% (from 22 MPa to 19 MPa). One of the negative aspects of crosslinking is the reduction in ductility of the polymer. Their study demonstrated that the strain to failure and ultimate tensile strength decreased progressively with increasing radiation dose. In spite of this, the wear resistance of the polymer benefited by increased crosslink density. The wear rate of the CISM-treated polymer decreased with increasing radiation dose and asymptotically reached a plateau value of immeasurable wear above 150 kGy. The effect of dose level on the wear behavior of the polymer is shown in Fiig. 6. The wear rate as measured on a bi-directional pin-on-disc machine decreased as a function of increasing radiation dose and displayed a transition from a 'wear' to 'no-wear' region at around 150 kGy. The authors also reported on the hip joint simulator wear behavior of CISM treated UHMWPE (150 kGy). At the end of 12 million cycles (simulating 12 years of use) the authors reported no measurable wear in the CISM-treated UHMWPE components. In contrast, the wear in the conventional UHMWPE components measured 207 mg (17 mg/million cycles) on average.

The above studies indicate that crosslinking has the potential to improve significantly the wear resistance of UHMWPE under lubricated conditions that either occur naturally in the human joint or are replicated in hip joint simulators. It should be noted that crosslinking is not without drawbacks. While simulator stud-

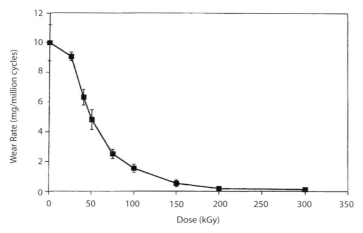

Fig. 6. Plot of wear rate of UHMWPE measured on a bi-directional pin-on-disc machine as a function of radiation dose. The plot displays a transition from a 'wear' to 'no-wear' region at around 150 kGy [95]

ies look quite promising, the crosslinked polyethylene suffers from inferior fatigue crack propagation resistance [96, 97]. Fatigue resistance is clearly important in total joint replacements as cyclic contact stresses can range from 10 MPa in tension to −40 MPa in compression at the articulating surface [98]. Recent studies have shown that the resistance to crack propagation is decreased with crosslinking whether it is achieved with chemical or ionizing methods [96]. Moreover, the stress intensity range required to initiate the growth of flaws is found to decrease with increasing radiation dose or crosslink density [97]. Figure 7 illustrates the detrimental effect of crosslink dose on the crack propagation resistance of UH-MWPE. Clearly this indicates that crosslinking must be optimized in order to improve wear resistance and to retain toughness within the polymer. Recent work by Klapperich et al. [99] has shown that low temperature plasma may be a viable technique to generate improved wear resistance while maintaining the bulk properties of the polymer. This result is depicted in Fig. 8. This technique can be used to control profiles of crosslinking near the articulating surface and to improve the biocompatibility of the polymer [100].

The developments of innovative bearing surfaces made of UHMWPE with optimized crosslink densities may provide improved in-vivo wear resistance and long-term structural stability. Such improvements are essential in avoiding periprosthetic osteolysis, the leading cause of implant failure, and may ultimately extend the life of total joint replacements. Such material improvements have the potential to make important contributions to the field of total joint arthroplasty.

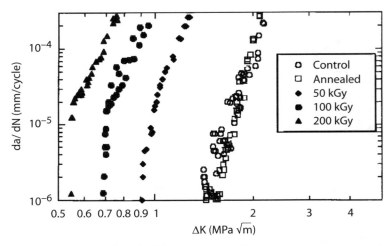

Fig. 7. Plot of crack propagation, da/dN, as a function of stress intensity range or crack driving force, ΔK for UHMWPE crosslinked over a range of radiation dosages. The plot illustrates the detrimental effect of crosslink dose on the crack propagation resistance of UHMWPE [97]

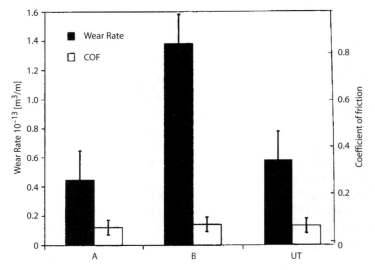

Fig. 8. Plot showing the wear rate of UHMWPE exposed to a range of low temperature plasma treatments [99]

4.3
The Effects of Ionizing Radiation on Acrylic Based Bone Cements

In cemented total joint arthroplasties, acrylic bone cement functions as the primary load bearing material used to transfer loads from the implant to the bone. Bone cement is formed from an exothermic reaction of benzoyl peroxide initiator present in polymethylmethacrylate powder (PMMA) and N,N-dimethyl-p-toluidene in methylmethacrylate monomer liquid (MMA), resulting in polymerization of PMMA to form a solid cement matrix. The in vivo integrity and performance of bone cement is necessary for longevity of orthopedic implants, because it is believed that mechanical failure of the bone cement layer can lead to aseptic loosening of the implant [101].

Many pre-surgical variables are expected to contribute to mechanical performance. Similar to UHMWPE, sterilization methods have been shown to affect porosity and molecular weight [102–104]. As with any implantable biomaterial, bone cement must be sterilized prior to its introduction in the body. In standard practice, EtO or gamma irradiation is used to sterilize the powder constituents while the liquid monomer is typically sterilized by membrane filtration. PMMA undergoes chain scission but not crosslinking when irradiated, leading to a reduction in its molecular weight [102–104]. This reduction has serious consequences since the molecular weight of a polymer is known to affect directly its mechanical properties [105].

In a recent study by Graham et al. [106] the relative effects of both sterilization method and vacuum-mixing techniques on the fatigue and fracture properties of acrylic bone cement were characterized. Five sterilization groups were examined: EtO, 2.5 Mrad γ radiation, 5 Mrad γ radiation, 10 Mrad γ radiation, and an unsterilized control group. The EtO and 2.5 Mrad groups provided clinically-relevant sterilization techniques while higher radiation doses enabled molecular weight degradation to be examined. Sterilization via gamma radiation resulted in large decreases in molecular weight. Table 3 summarizes the molecular weight of the PMMA as a function of radiation dose.

Table 3. Weight-averaged molecular weights of Palacos R brand PMMA powders and cements for all sterilization groups [106]. Values are reported as mean (standard deviation)

	Control	EtO	2.5 MRad	5 MRad	10 MRad
M_w powder	856,094 (17,990)	850,576 (9294)	310,466 (8797)	188,490 (6178)	105,677 (–)
M_w cement	768,449 (9300)	743,616 (22,389)	302,965 (4911)	212,229 (5027)	

Table 4. Fracture toughness of hand and vacuum mixed bone cement under different sterilization conditions. Values are reported as mean (standard deviation). n indicates number of samples in each group [106]

Fracture toughness K_{Ic} (MPa $\sqrt{m}$)	Control	EtO	2.5 MRad	5 MRad	10 MRad
Hand mixed samples	1.7642 (0.1097) n=20	1.8199 (0.0805) n=19	1.7367 (0.1049) n=15	1.7458 (0.1046) n=16	– –
Vacuum mixed samples	1.9502 (0.1744) n=13	2.0003 (0.2549 n=16	1.8197 (0.0581) n=13	– –	1.6883 (0.0856) n=14

Table 5. Number of cycles to failure for hand and vacuum mixed bone cement sterilized using
gamma radiation or EtO gas. Values are reported as mean (standard deviation). n indicates
number of samples in each group

No. of cycles to failure	EtO	2.5 MRad	5 MRad	10 MRad
Hand mixed samples	763 (369) n=12	838 (664) n=9	959 (442) n=8	– –
Vacuum mixed samples	2040 (2234) n=8	1624 (434) n=5	– –	548 (172) n=8

The results of the fracture toughness tests are summarized in Table 4. For cements
that were vacuum-mixed, both the gamma irradiation treatments (2.5 MRad and
10 MRad) caused statistically significant decreases in fracture toughness from control values while EtO sterilization caused no significant change. Hand mixing, owing
to higher degree of porosity, provided no statistically significant difference amongst
the sterilization methods. These findings indicate that ionizing radiation results in
deterioration of both the molecular weight and the fracture toughness, while EtO
sterilization causes neither of these changes. Fatigue properties were also affected by
choice of sterilization method (Table 5). The severe molecular weight degradation
associated with 10 MRad γ radiation resulted in a significantly lower fatigue life than
either the EtO or the 2.5 MRad γ radiation groups. The results of this study indicate
that the decrease in molecular weight caused a significant decrease in both fracture
toughness and fatigue resistance. This study did not address issues of aging, on the
shelf or in vivo, but it is important to note that gamma irradiation also leads to the
generation of free radicals which may have long-term harmful effects. Studies using
electron paramagnetic resonance spectroscopy have demonstrated that these radia-
tion-generated free radicals in bone cement are long-lived and are not annihilated
in the polymerization process [107]. The role of these free radicals on the in vivo
properties of bone cement is not well known: it has been conjectured that free radicals within the cement may contribute to its in vivo degradation. A recent study on

bone cement retrieved from patients clearly confirms a greater degradation in the molecular weight of gamma-irradiated bone cement than non-irradiated cement with increased time in vivo [108]. This finding suggests that the deterioration of mechanical properties of gamma-sterilized cement seen in this study could continue to worsen with in vivo use.

5
Summary

Medical polymers are structurally altered when exposed to ionizing radiation. Certain classes of polymers are susceptible to chain scission mechanisms while others undergo crosslinking processes. The atmosphere in which the polymer is radiated complicates these generalizations. Because these devices must be sterilized prior to implantation their long-term structural evolution owing to the effects of ionizing radiation coupled with environment must be understood. This work summarizes the wide-ranging effects of ionizing radiation on the mechanical and structural properties of medical polymers.

References

1. Matthews IP, Gibson C, Samuel AH (1994) Biomaterials 15:191
2. Zimmerli W, Woldvogel FA, Vandaux P, Nydegger UE (1982) Curr Prob Surg 146:487
3. Shinozaki T, Deane RS, Mazuzan JE, Hamel AJ, Hazelton D (1983) J Am Med Assoc 249:223
4. Leake ES, Glristina AG, Wright MJ (1982) J Clin Microbiol 13:320
5. Stone HH, Fabian TC, Turkeson ML, Jurkiwicz MJ (1981) Ann Surg 193:612
6. Goldman M, Pruitt L (1998) J Biomed Matls Res 40:378
7. Shalaby S (1993) Radiation effects on polymers of biomedical significance. In: Reichmanis E, Frnk CW, O'Donnell HO (eds) Irradiation of polymeric materials. ACS, Washington, p 315
8. Gaughran ERL, Goudie AJ (eds) (1978) Sterilization of medical products by ionizing radiation. Multisience Publ, Montreal
9. Harris LE, Skopek AJ (eds) (1985) Sterilization of medical products. Johnson and Johnson, New South Wales
10. Reichmanis E, Frank CW, O'Donnell HO (eds) (1993) Irradiation of polymeric materials. ACS, Washington
11. Clough RL, Shalaby SW (eds) (1991) Radiation effects on polymers. ACS, Washington
12. Clegg DW, Collyer (eds) (1991) Irradiation effects on polymers. Elsevier, London
13. Dole M (ed) (1972) Radiation chemistry of macromolecules, vol I. Academic Press, New York
14. Dole M (ed) (1973) Radiation chemistry of macromolecules, vol II. Academic Press, New York
15. Chapiro A (1962) Radiation chemistry of polymeric systems. Interscience, New York
16. Charlesby A (1960) Atomic radiation and polymers. Pergamon Press, London
17. Singh A, Silverman J (eds) (1992) Radiation processing of polymers. Hanser, Munich
18. Parkinson WW (1987) Radiation resistant polymers in EPST. Oak Ridge National Lab, Oak Ridge
19. Brynjolfsson A (1974) Sterilization by ionizing radiation:145. Multiscience, Montreal

20. Harrod RA (1977) Radiat Phys Chem 9:91
21. Eymery R (1974) Sterilization by ionizing radiation:84. Multiscience, Montreal
22. McLaughlin WL (1978) Conference on national and international standardization of dosimetry. NIST, Gaithersburg
23. Tomita K, Sugimoto S, (1977) Radiat Phys Chem 9:576
24. Burton M (1952) Disc Faraday Soc 12:317
25. Lyons BJ, Fox AS (1968) J Polym Sci C 21:159
26. Moore PW(1993) In: Frank CW, O'Donnell HO (eds) Irradiation of polymeric materials, vol 2. ACS, Washington, p 9
27. Charlesby A (1954) Proc R Soc London A222:542
28. Saito O (1959) J Phys Soc Japan 13:798
29. Saito O (1972) In: Dole M (ed) The radiation chemistry of macromolecules, vol 11. Academic Press New York, p 223
30. Clough RL (1980) J Am Chem Soc 102:5242
31. Wasserman H, Murray R (1979) Singlet oxygen. Academic Press, NY
32. Clough RL, Gillen KT, Quintana CA (1985) J Polym Sci Polym Chem Ed 23:359
33. Bowmer T, Cowen L, O'Donnell, Winzor D (1979) J Appl Polym Sci 24:425
34. Matsuo H, Dole M (1959) J Phys Chem 63:395
35. Wundrich K (1973) J Polym Sci Polym Phys Ed 11:1293
36. Clough RL, Gillen KT, Campan JL, Gaussens G, Schonbacher H, Seguchi T, Wilski H, Machi S (1984) Nucl Safety 25:198
37. Kusy RP, Turner DT (1971) Macromolecules 4:337
38. Bhateja SK, Andrews EH, Young RJ (1983) J Polym Sci Polym Phys Ed 21:523
39. Charlesby A (1991) In: Clegg DW, Collyer (eds) Irradiation effects on polymers. Elsevier, London
40. Ratner B, Hoffman AS, Schoen FJ, Lemons JE (1996) Introduction to biomaterials science. Academic Press, New York
41. Park JB (1979) Biomaterials: an introduction. Plenum Press, New York
42. Keller A (1981) In: Basset DC (ed) Developments in crystalline polymers. Applied Science, London, p 37
43. Bhateja SK (1983) J Appl Polym Sci 28:861
44. Salovey R, Bassett DC (1964) J Appl Phys 35:3216
45. Shinde A, Salovey R (1985) J Polym Sci 23:1681
46. Klein PG, Gonzalez-Orozco JA, Ward IM (1991) Polymer 32:10
47. Chen CJ, Boose DC, Yeh GSY (1991) Colloid Polym Sci 269:469
48. Dole M, Milner DC, Williams TF (1958) J Am Chem Soc 80:1580
49. Dijkstra DJ, Hoogsteen W, Pennings AJ (1989) Polymer 30:866
50. Grobbelaar CJ, Du Plessis TA, Marais F (1978) J Bone Joint Surg 60B:374
51. Wroblewski BM, Siney PD, Dowson D, Collins SN (1996) J Bone Joint Surg 78B:280
52. Shen FW, McKellop HA, Salovey R (1995) J Poly Sci Poly Phys 24:1063
53. Oonishi H (1995) Orthop Surg Trauma 38:1255
54. McKellop H, Shen FW, Salovey R (1998) Trans Orth Res Soc 23:98
55. Premnath V, Harris WH, Jasty M, Merrill EW (1996) Biomaterials 17:1741
56. Williams D (1978) Stabilization and degradation of polymers. In: Allara (ed) Advances in chemistry 169:142
57. Dunn TS, Epperson BJ, Sugg HW, Stannett VT, Williams JL (1979) Radiat Phys Chem 14:625
58. Gavrila DE, Gosse B (1994) J Radioanal Nucl Chem 185:311
59. Hegazy EA, Seguchi T, Arakawa K, Machi S (1981) J Appl Polym Sci 26:1361
60. Pruitt L (2001) Encyclopedia of materials: science and technology. Williams DF (ed) Advances in polymer science. Elsevier Science Limited, Oxford
61. Lyons BJ (1995) Radiat Phys Chem 45:159
62. Uscold RE (1984) Appl Polym Sci 29:1335

63. Licht WR, Kline DE (1964) J Polym Sci A 2:4673
64. Harada K, Kogure O, Murase K (1982) IEEE Trans Electron Dev 29:518
65. Dickens B, Martin JW, Waksman D (1984) Polymer 25:107
66. Lyons BJ, Glover LC (1991) Radiat Phys Chem 37:93
67. Zamore A (2001) http//nasatech.com
68. Sanford WM, Lilly WB, Moore WC (1996) Trans 42nd Orthop Res Soc 21:23
69. Choudhury M, Hutchings IM (1997) Wear 203:335
70. Rimnac CM, Burstein AH, Carr JM (1994) J Appl Biomater 5:17
71. Pascaud RS, Evans WT, McCulagh PJ, FitzPatrick D (1995) Soc Biomater 18:809
72. Rimnac CM, Klein RW, Burstein A, Wight TM, Santner T (1994) Trans Orth Res Soc 19:175
73. Goldman M, Gronsky R, Long GG, Pruitt L (1998) Polymer Degr Stab 62:97
74. McCord JM (1983) Surgery 94:412
75. Frank MR, Smith JJ, Bacon RC (1964) J Polym Sci 13:535
76. Kurtz SM, Muratoglu OK, Evans M, Ediden AA (1999) Biomaterials 20:1659
77. Kurtz SM, Rimnac CM, Bartel DL (1997) J Orthop Res 15:57
78. Besong AA, Hailey JL, Ingham E (1997) Biomed Mater Eng 7:59
79. Currier BH, Currier JH, Collier JP, Mayor MB, Scott RD (1997) Clin Orthop 342:111
80. Sun DC, Stark C, Dumbleton JH (1994) Polymer 35:969
81. Sun DC, Smidig G, Stark C, Dumbleton JH (1996) Trans Orthop Res Soc 21:493
82. Sanford WM, Saum KA (1995) Trans Orthop Res Soc 20:119
83. Crane D, Kurtz SM, Pruitt L, Edidin AA (1998) Trans Soc Biomater 21:503
84. Frank MR, Smith JJ, Bacon RC(1954) J Polym Sci 13:535
85. Bartell DL, Rimnac CM, Wright TM (1991) In: Goldberg VM (ed) Controversies of total knee arthroplasty. Raven Press, New York, p 61
86. Baker D, Hastings R, Pruitt L (2000) Polymer 41:795
87. Jasty M, Bragdon CR, O'Connor DO (1997) Trans Orthop Res Soc 43:785
88. McKellop H (1996) Gamma radiation crosslinked UHMWPE. Harvard Hip Course, Boston
89. Oonishi H, Takayama Y, Tsuji E (1992) Rad Phys Chem 39:495
90. McKellop H, Shen FW, Salovey R (1998) Trans Orthop Res Soc 23:98
91. Edidin AA, Jewett CW, Foulds JR, Kurtz SM (1998) Trans Soc Biomater 21:220
92. Wang A, Polineni VK, Essner A (1997) Trans Soc Biomater 20:394
93. Oonishi H, Takayama Y, Tsuji E (1992) Radiat Phys Chem 39:495
94. Jasty M, Bragdon CR, O'Connor DO (1997) Trans Orthop Res Soc 43:785
95. Muratoglu OK, Bragdon CR, O'Connor DO (1997) Soc Biomater 20:74
96. Baker D, Hastings R, Pruitt L (1999) J Biomed Mat Res 46:573
97. Baker D, Bellare A, Pruitt L (2001) J Mat Sci Lett
98. Bartell DL, Bicknell VL, Wright TM (1986) J Bone Jnt Surg 68:1041
99. Klapperich C, Komvopoulos K, Pruitt L (1999) In: Biomedical materials for drug delivery, medical implants, and tissue engineering, p 331
100. Klapperich C (2000) Ph D Thesis, UC Berkeley
101. Davies J, Harris W (1989) Soc Biomat 89
102. Harper EP, Braden M, Bonfield W, Dingeldein H, Wahlig H (1997) J Mater Sci Mater Med 8:849
103. Lewis G, Mladsi S (1998) Biomaterials 19:117
104. Goldman M, Pelletier B, Muller S, Ries M, Pruitt L (1998) Trans Orthop Res Soc 217
105. Hertzberg RW (1983) Deformation and fracture mechanics of engineering materials. Wiley, New York
106. Graham J, Ries M, Pruitt L, Gundiah N (2000) J Arthroplasty 15:1028
107. Park JB, Turner RC, Atkins PE (1980) Biomater Med Dev Artif Organs 8:23
108. Pelletier B, Hughes K, Gundiah N, Muller S, Pruitt L, Ries M (1999) Trans Orthop Res Soc 513

Effects of Ion Radiation on Cells and Tissues

M. Scholz

GSI/Biophysik, Planckstrasse 1, 64291 Darmstadt.
E-mail: m.scholz@gsi.de

Studies of the biological action of ionizing radiation on cells and tissues are of interest for applications in radiotherapy as well as radiation protection. In general, the biological response to ion beam irradiation considerably differs from that to conventional photon beam irradiation. This difference for example constitutes one of the major rationales for the application of ion beams in tumor therapy.

The review first summarizes the biological effects of conventional photon radiation to cells and tissues and describes the basic techniques for their quantification. The essential physical characteristics of ion beams, which are relevant to understand their particular biological effects, are then described. The systematics of biological effects of ion beams is presented first in detail for cellular systems, complemented by a brief overview of the action on more complex biological systems like tissues. Furthermore, aspects of biophysical modeling of radiation effects are discussed, and a model recently developed within the framework of charged particle beam radiotherapy is presented in more detail. In the final part, a brief description of the important technical and biological aspects of ion beams in tumor therapy is given.

Keywords: Ionizing radiation, Ion beam, Biological effect, Tissue, Tumor therapy

1	**Introduction**	97
2	**General Aspects of Radiation Damage to Cells and Tissues**	99
2.1	Organization of Cells and Tissues	99
2.2	Cellular Effects of Radiation	101
2.3	Quantifying Radiation Effects	103
2.3.1	Strand Breaks	103
2.3.2	Chromosome Aberrations	106
2.3.3	Cell Survival	108
2.3.4	Modification of Radiosensitivity	110
2.3.5	Molecular Techniques	111
2.4	Radiation Response on the Tissue Level	112
3	**Physical Characteristics of Ion Beams**	114
3.1	Microscopic Features	114

Advances in Polymer Science, Vol. 162
© Springer-Verlag Berlin Heidelberg 2003

3.2   Macroscopic Features 117
3.3   Neutron Beams 119

4   Biological Effects of Ion Irradiation 120

4.1   The Relative Biological Effectiveness (RBE): Definition 120
4.2   Systematics of RBE 121
4.2.1   Dose Dependence 121
4.2.2   Energy/LET Dependence 122
4.2.3   Particle Dependence 123
4.2.4   Cell Type Dependence 124
4.3   Very Heavy Particles 126
4.4   Inactivation Cross Section 129
4.5   Systematics of Inactivation Cross Sections 129
4.6   Oxygen Effect 132
4.7   The Role of Increased Ionization Density 132
4.7.1   Double Strand Break Induction and Rejoining 133
4.7.2   Chromosome Aberrations 134
4.7.3   Fractionated Irradiation 135
4.7.4   Direct Visualization of Localized Damage 136
4.8   Bystander Effects 136
4.9   Tissue Effects 138
4.9.1   Normal Tissues 138
4.9.2   Tumor Tissue 139

5   Models of Biological Action of Heavy Charged Particles 140

5.1   Theory of Dual Radiation Action 140
5.2   Cluster Models 141
5.3   Amorphous Track Structure Models 142
5.4   Influence of Target Structure 145

6   Application of Charged Particle Beams in Tumor Therapy 145

6.1   Technical Realization 146
6.2   Radiobiological Considerations 147
6.2.1   Optimal Ion Species 147
6.2.2   Depth Dependence of RBE 148
6.3   Clinical Aspects 150

7   Summary and Perspective 151

References 152

List of Abbreviations

bp	Base pair
CHO	Chinese hamster ovary cell line
DSB	Double strand break
LEM	Local effect model
LET	Linear energy transfer
OER	Oxygen enhancement ratio
p21	Protein involved in cell cycle regulation
RBE	Relative biological effectiveness
SSB	Single strand break
TDRA	Theory of dual radiation action
V79	Chinese hamster lung fibroblast cell line
XRS	X-ray sensitive mutant cell line

1
Introduction

To study radiation effects on cells and tissues is of particular interest for several fields of application such as, e.g., radiation protection or radiotherapy of cancer. The major concern of radiation protection is to understand the *induction of malignancies*, in particular cancers, by radiation. Although not yet fully understood, it is assumed that cancer induction is initiated by small changes of genetic information in individual cells (mutations) [1]. At the other extreme, radiation is also applied to *cure malignancies* like cancer, explicitly utilizing the tissue damaging properties of ionizing radiation to kill the tumor cells. However, dose deposition to a tumor is in general connected with the simultaneous deposition of dose in the surrounding normal tissue. Therefore, application of radiation in tumor therapy always represents a compromise between maximal damage to the tumor tissue and minimal side effects to the surrounding healthy tissue. These side effects include acute reactions observed within days or weeks after therapy, late reactions typically occurring months or years after therapy and radiation induced secondary cancers, which will appear up to 20 or 30 years after treatment.

About one-third of all localized tumors, which contribute to about 60% of all cancers, are cured by radiotherapy alone or by radiotherapy in combination with surgery [2]. For another one-third of the localized tumors, however, a failure of local control is observed, and this group is expected to benefit from the development of techniques allowing a better conformation of dose to the tumor. The compromise between tumor cure and minimal side effects is largely determined by the physical properties of the radiation type under consideration, namely the depth dose profile. The depth dose profile defines the ratio of the dose delivered to the tumor and the dose delivered to the healthy tissue.

Ion beams are characterized by an advantageous depth dose profile, which makes them in particular suitable for applications in radiotherapy. These advantages were already recognized in the 1940s [3]. However, due to the considerably higher efforts to produce an ion beam suitable for therapy compared to the production of conventional electron and photon beams, their application in radiotherapy was restricted to a few places worldwide [4]. But meanwhile there is significantly growing interest in ion beam radiotherapy (for the most recent compilation of facilities, see the 'Particles Newsletter'[1]). The successful installation of clinically based facilities as well as pilot projects using most advanced accelerator and beam delivery techniques have substantially contributed to this growing interest.

The characteristic radiobiological properties of ion beams represent a further advantage with respect to their application in tumor therapy. In general, ion beams show a higher effectiveness of cell killing compared to photon beams in particular at the end or their penetration path [5–7]. Using appropriate beam delivery techniques, the region of increased effectiveness can thus essentially be focused to the tumor volume.

The precise knowledge of the biological action of charged particle radiation is also relevant for radiation protection purposes, since also for other biological effects like cell transformation an increased effectiveness of charged particle beams is observed [8]. For example, this is crucial for risk estimates for miners, related to the inhalation of α-emitting isotopes. Furthermore, it is also a relevant issue in radiation protection in space; heavy charged particles are responsible for a significant fraction of radiation damage from cosmic rays [9].

When discussing the biological action of radiation – whether due to photon, electron or ion beams – on living cells and tissue, several differences to the action on non-living material have to be considered. Obviously, the most important difference is the capability of living cells to actively *process* the damage induced by radiation, so that we have to distinguish the primary induced damage from the residual damage, which eventually is responsible for the observable effect.

Processing of damage includes a whole spectrum of possible reactions, *repair* of the damage being the most important one. Since, however, the fidelity of the repair process is not perfect and depends on the particular type of damage induced, there is a certain possibility of misrepair or other modifications of the primary damage. Processing of damage takes time, and in combination with other kinetic effects it thus implies a pronounced time dependence of biological effects, spanning a huge time scale from minutes to years until the primary induced damage is actually converted into a visible or detectable biological response.

Furthermore, the radiation response of tissues is characterized by a complex interplay between different cell types, each of them showing an individual response to radiation damage.

---

[1] http://neurosurgery.mgh.harvard.edu/hcl/ptles.htm

Section 2 briefly summarizes the most important characteristics of typical cells and tissues and their response to conventional photon radiation. This will facilitate understanding of the particular radiobiological properties of ion beams. It also includes an overview of the most important systems and techniques used to quantify radiation effects on cells and tissues. The biologically relevant physical characteristics of ion beams as compared to photon beams will be introduced in Sect. 3. The description of the particular biological effects of ion beams will then be presented in Sect. 4. Section 5 introduces aspects of modeling of the radiobiological properties of ion beam radiation. As an example for the application of ion beams, Sect. 6 describes the use of ion beams in tumor therapy. Finally, Sect. 7 summarizes the most important aspects and gives some remarks on future perspectives of ion beam applications.

2
General Aspects of Radiation Damage to Cells and Tissues

This section will give a brief overview of the organization of cells and tissues from the perspective of radiobiology. It first describes the general characteristics of a typical cell and how tissues are composed of cells. In the second part the essential cellular effects of radiation will be summarized, followed by an introduction to the basic experimental methods and procedures which are used to quantify radiation effects on the cellular level. The final part then presents a description of the typical radiation response on the tissue level.

2.1
Organization of Cells and Tissues

A single cell represents the smallest functional unit of any complex organized tissue. In general, within a single cell two clearly separated compartments can be distinguished visually and functionally: the cell nucleus and the cytoplasm. The cell nucleus contains the genetic information in the form of a large macromolecule, the DNA. In combination with additional proteins, secondary, tertiary, and higher order structures are built, resulting in a condensed structure of the DNA molecule.

Within the cytoplasm, further substructures (organelles) can be distinguished. These comprise, e.g., the mitochondria (responsible for the energy production), the endoplasmic reticulum in combination with the ribosomes (which are involved in the assembly of proteins), and the Golgi apparatus (involved in further processing and transport of macromolecules within the cell and out of the cell). All the compartments are separated by membranes, which allow concentration gradients of certain types of ions or molecules to remain. This is also true for the outer cell membrane, separating the inner cell volume from the environment.

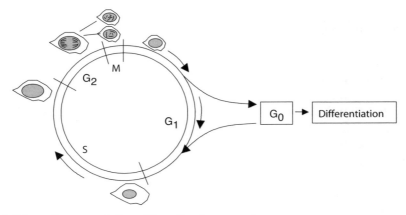

Fig. 1. Schematic representation of the cell cycle progression

A typical characteristic of many cells is their ability to grow and to produce two identical daughter cells by cell division. This division requires the exact duplication of the DNA contained in the cell nucleus, and the precise distribution of each of the two copies into the daughter cells. In general, proliferating cells in tissue as well as under laboratory conditions show a very regular division cycle, which is schematically shown in Fig. 1. Beginning with a cell that was just produced by division of a predecessor, it starts with a preparation phase, which is necessary to initiate the DNA replication (G_1-phase).

It is followed by the replication or synthesis of DNA (S-phase), and before cell division takes place a second preparation phase (G_2-phase) is required. During the short interval of mitosis (M-phase), the DNA is packed in an extremely condensed form, microscopically visible as chromosomes, which are then symmetrically distributed to the two daughter cells. The total time for a complete division cycle of typical mammalian cells under laboratory conditions is in the order of 12–24 h.

The progress through the cell cycle and the DNA synthesis are highly organized and controlled processes [10]. Proliferation, e.g., depends on the environmental conditions and the integrity of the DNA molecule. Under certain circumstances, cells can leave the regular cycle and stay in a resting phase (G_0-phase), but upon appropriate stimuli are still able to reenter the normal cycle.

The control of proliferation is particularly important with respect to complex organisms consisting of millions of individual cells, where a continuous, uncontrolled growth would be incompatible with the required complex interplay of different cell types and organs.

Most organs and tissues are composed of different cell types. These comprise stem cells, which are capable in principle of unlimited division and renewal, as well as their descendants, passing through a chain of differentiation and maturation steps. 'Differentiation' describes the transition of a proliferating, growing cell to a cell which has lost its proliferating capacity, but remains in a status where it can fulfill certain specialized functions. This transition is indicated in Fig. 1 by the arrow to G_0-phase/differentiation, symbolizing the departure from the regular cell cycle. Stem cells as the origin of functional cells represent the basic unit with respect to recovery of the tissue from externally induced injury.

The compartmentalization of tissues is maintained by a complex network of signaling and interaction between the different cells and cell types. Signaling between cells can be achieved in principally two different ways:

- By exchange of small signaling molecules through particular channels ('gap junctions') connecting neighboring cells; this type of interaction requires direct cell-cell contacts [11].
- By diffusion-controlled exchange of molecules through the intercellular space, which are recognized by the target cells through receptors on the outer cell membrane.

The relative contribution of both of these pathways depends on the particular tissue under consideration. The information exchange between cells plays a key role for the so-called 'bystander effect', which describes the fact that even cells not directly damaged by radiation can be affected indirectly by signals received from directly damaged cells [12–14].

2.2
Cellular Effects of Radiation

Starting from the structural complexity of a single cell, the question arises which compartment is most sensitive to radiation and can thus be expected to be responsible for the observable response of a cell to radiation? Experimental results using viruses, bacteria, yeast, and mammalian cells have demonstrated a correlation between the radiosensitivity and the DNA content, at least for groups of biologically similar objects: the higher the amount of DNA, the more sensitive the object (overview in [15]). These results already suggested that DNA plays a key role in the response to radiation. This hypothesis has been proven also more directly for mammalian cells [16]. The experiments revealed, that energy deposition in the nucleus is by far more efficient to produce biological damage, compared to the case where similar amounts of energy are deposited to the cytoplasm only. Cells were shown to respond to energy deposits in the nucleus at a level approximately 100 times lower than that required to detect similar biological effects, when only the cytoplasm is irradiated. Several other experiments also support the view that the DNA molecule represents the critical target for radiation effects in cells. However, there is increasing evidence in the last few years that DNA damage is not necessarily a

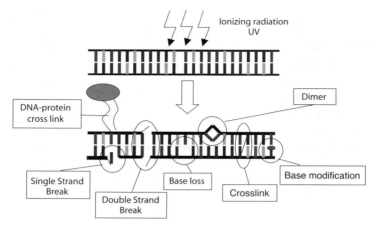

Fig. 2. Radiation induced DNA damage. For clarity, the DNA double helix is drawn as a flat, ladder-like structure

prerequisite for the induction of biologically relevant effects (this point will be discussed in more detail in Sect. 4.8).

DNA damage can be induced by radiation in two different ways. On the one hand, radiation leads to ionizing events in the DNA molecule itself, subsequently leading to breakage of molecular bonds and disruption of one or both strands of the DNA. These events are termed 'direct effect'. On the other hand, radiation leads to the production of, e.g., highly reactive OH-radicals by radiolysis of the water molecules surrounding the DNA molecule. These radicals are able to migrate over distances of a few nanometers during their lifetime and are thus capable of damaging the DNA molecule, even if produced at a certain distance. This action is termed 'indirect effect' [15].

Figure 2 summarizes the major types of DNA damage induced by ionizing radiation, whether by the direct or by the indirect effect. It has to be taken into account, however, that these types of lesions do not necessarily occur separately, but instead, depending on the dose level, combinations of different types occurring in close vicinity can lead to more complex lesions. Since the information on both strands of the DNA molecule is complementary, all injuries affecting only one side of the DNA double strand can potentially be easily repaired by using the information on the intact strand as a template.

Therefore, double strand breaks (DSB) are generally considered as the critical event for the induction of lethal lesions [17–19]. Table 1 summarizes the incidence of several types of lesions after application of 1 Gy to a typical cell. These numbers, however, should only illustrate an order of magnitude; there can be considerable variations from cell type to cell type.

Table 1. Approximate yields of DNA damage per Gy per cell (estimated from [20–22])

SSB	1000
DSB	30–40
DNA-protein crosslinks	50
Complex damage (SSB+base lesion)	60

As mentioned already, higher organisms like, e.g., mammalian cells are in general able to recognize and to repair damage to DNA at least to a certain extent [23, 24]. The efficiency of these repair processes depends on the complexity of the damage induced. For example, single strand breaks can be repaired comparatively easily, because this type of lesion resembles naturally occurring events during the replication cycle, e.g., when the double strand has to be opened on one strand to allow the access of replication proteins to the DNA. The protein machinery of the cell is well prepared to handle these events. With increasing complexity, however, damage becomes more difficult to repair, and this might enhance the probability that the repair process cannot be accomplished correctly, leaving a partially repaired or modified DNA molecule [25, 26].

The investigation of cellular repair pathways is an important field of current biological research, and these processes are by far not yet fully understood. Most studies have been performed using relatively simple biological objects like, e.g., bacteria and yeast cells, but for the greatest part of pathways it could be shown that analogous mechanisms exist also in more complex organisms like, e.g., mammalian cells.

2.3
Quantifying Radiation Effects

Comparison of the effectiveness of different radiation types requires an accurate quantification of radiation effects. This section describes the basic experimental procedures which are used in classical radiobiological studies, covering different biological levels including DNA strand breaks, chromosome aberrations, and cell death as defined by the loss of proliferative capacity.

2.3.1
Strand Breaks

Direct investigation of DNA strand break induction is often performed using particularly simple, small DNA molecules from viruses or plasmids. These systems have the advantage that induction of SSB or DSB leads to characteristic topological changes of the molecule, which allow one to identify unambiguously the fraction of molecules containing no damage at all, a single strand break, or a double strand

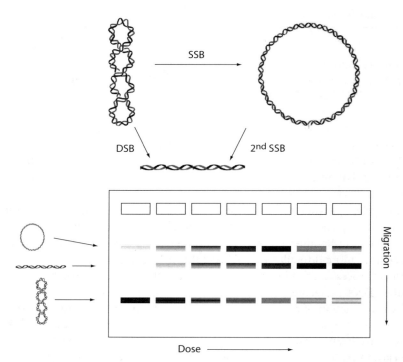

Fig. 3. *Top*: Induction of strand breaks in viral and plasmid DNA. The undamaged (native) form has a supercoiled structure. Induction of a single strand break leads to an open ring form, whereas by induction of a double strand break the molecule is transformed into a linearized form. The linearized form can also result from induction of two single strand breaks in close vicinity (Courtesy S. Brons). *Bottom*: The three conformations are characterized by different migration velocities in the gel.

break (see Fig. 3a). The various molecule conformations are characterized by different migrations velocities in gel electrophoresis, leading to characteristic bands as shown schematically in Fig. 3b.

Studying strand breaks in mammalian cells is more demanding. Whereas the size of viral or plasmid DNA is in the order of some thousand basepairs (bp), mammalian cells typically contain in the order of 3×10^9 bp. Inducing approximately 350 DSB by irradiation with 10 Gy – which is in the order of typical doses applied in cell experiments – roughly corresponds to 1 DSB/10^7 bp. This is much lower than the value of approximately 1 DSB/10^4 bp which is the order of magnitude for investigations using plasmid DNA. In addition, induction of DSB in mammalian DNA does not lead to defined conformational changes of the DNA molecule as in plasmid DNA. Therefore, the number of strand breaks has to be estimated from the production of DNA fragments. According to the random distribution of energy deposition events (see below), a random distribution of breaks within the genome can be expected in a first approximation, and fragment sizes will cover a correspondingly broad spectrum. Small fragments (<6–9 Megabasepairs) can be separated from large fragments and undamaged DNA by gel electrophoresis. According to the broad spectrum of fragment sizes, fragmented DNA appears as a broad smear (see Fig. 4). From the fraction of small sized DNA, the number of DSB can then be estimated using appropriate calibration procedures.

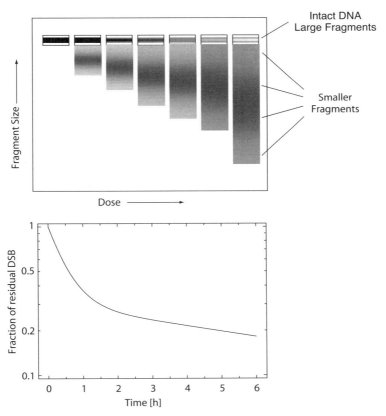

Fig. 4. *Top*: Gelelectrophoretic detection of double strand breaks in cellular DNA. Intact DNA as well as large fragments are retained in the plugs, whereas smaller fragments migrate in the gel. The total amount of DNA detected in the broad smear below the plug is a measure of the number of induced DSB; the width of the distribution depends on the fragment length distribution. *Bottom*: Rejoining of DSB can be detected by measuring the kinetics of the fraction of small fragments as a function of time after irradiation. During the incubation interval, strand breaks are rejoined and thus larger fragments are reconstituted, so that the fraction of residual DSB decreases with time.

Gel electrophoretic methods are also suitable to study the processing of DNA strand breaks such as, e.g., rejoining of the open ends by measuring the fraction of DNA fragments as a function of time after irradiation [27]. During incubation, strand breaks can be repaired or rejoined, so that larger size DNA fragments are reconstituted from smaller fragments. Repair and rejoining have to be distinguished here, since with the methods described here the restitution of DNA length can only be detected within certain limits, but the loss of very small fragments of DNA during the rejoining process cannot be detected. Figure 4b gives a typical example of such a rejoining curve after photon irradiation. Decay of the fraction of damaged DNA obeys an exponential law with a fast and a slow component. Rejoining of strand breaks is not always complete and, depending on radiation type and cell line, a certain fraction of residual damage can be observed even after long intervals of incubation. This residual damage is frequently used as indicator for the lethality of a given radiation type [25, 28].

2.3.2
Chromosome Aberrations

From the rejoining of DNA DSB no conclusions about the *fidelity* of the rejoining process can be drawn in general. A more detailed study of misrejoining and misrepair processes can be performed based on the analysis of chromosome aberrations. The advantages of using chromosome aberrations as indicators for radiation effects are that doses as low as 0.05 Gy can be detected, and that aberrations can be directly observed in the microscope [29].

Since the duration of mitosis, in which the DNA becomes extremely condensed and microscopically visible as chromosomes, is very short (20–30 min) compared to the complete cell cycle duration (typically 12–24 h), only a small fraction of cells can be found in mitosis at a given time interval. In order to increase the yield of mitotic cells, cell populations are treated using blocking agents like, e.g., colcemide, which prevent cells from completing mitosis, but do not affect the proliferation up to mitosis, so that with increasing time (typically 2 h) mitotic cells accumulate.

An example of a radiation induced chromosome alteration is shown in Fig. 5a; these 'dicentric' chromosomes are often used as marker for lethality. Figure 5b schematically explains how these aberrations can occur by misrejoining of two DSB produced in two adjacent chromosomes. The lethality of dicentric chromosomes is attributed to the fact that during cell division the correct segregation of these chromosomes and the accompanying acentric fragment is not guaranteed, so that genetic information is not symmetrically distributed between the daughter cells.

Although direct visibility of chromosome aberrations is a significant advantage, it has to be taken into account that not all types of aberrations become visible. For example, if approximately equally sized pieces of DNA are exchanged between chromosomes (which can happen if two DSB are produced in each of the two chromosomes, respectively), this exchange could not be detected by means of standard staining techniques as shown in Fig. 5. However, using more advanced techniques like fluorescence in situ hybridization (FISH), it is possible to stain selectively the DNA of one particular chromosome with a distinct fluorescence color or even to stain each of the chromosomes with a different color, so that any complex type of exchange of DNA chromosome pieces will become visible [30].

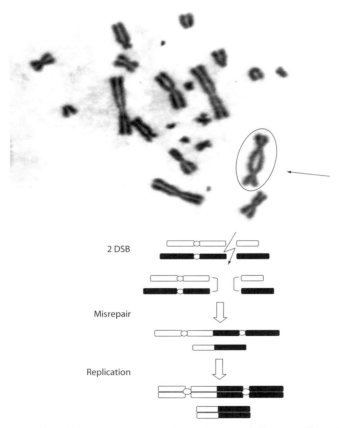

2 DSB

Misrepair

Replication

Fig. 5. Radiation induced chromosome aberration. The *arrow* indicates a dicentric chromosome. This aberration type is often used as marker for lethal lesions. (Courtesy S. Ritter). The schematic drawing below illustrates the mechanism leading to the formation of dicentric chromosomes by misrejoining of chromosome fragments resulting from two DSBs in two different chromosomes in close vicinity

A second disadvantage that has to be taken into account is that investigation of chromosomes requires the proliferation of cells up to mitosis. However, since radiation damage also affects the proliferation of cells, severely damaged cells can be expected not to reach mitosis, leading to an underestimation of radiation effects, particularly at high doses [31, 32]. This disadvantage can be partially circumvented by using special preparation techniques, where condensation of DNA into chromosomes is artificially induced also in the other cell cycle phases like, e.g., G_1– and G_2–phase ('premature chromosome condensation', PCC) [33, 34].

2.3.3
Cell Survival

Investigation of radiation induced cell death, defined as mitotic death in the sense of a complete loss of the proliferation capacity, is one of the most commonly used methods to study radiation effects on cells. As mentioned earlier, many cell types are characterized by regular cell division in 12–24 h intervals. Thus, according to the exponential growth, a single cell can produce thousands of daughter cells within a few days. If the cells are originally seeded in culture flasks at the appropriate low density, the daughter cells of each individual cell appear as clusters or 'colonies'. A cell is classified as 'survivor' if it is able to produce at least 50 daughter cells within a time interval of approximately 10–14 normal division cycles, i.e., 5–14 days; if less than 50 daughter cells are produced the cell is classified as dead or 'inactivated' [35]. The threshold of 50 cells is an empirically determined value and somewhat arbitrary; actually there is no clear-cut value defined because there is a smooth transition between cells producing no daughter cells at all and cells producing the maximum possible value of 2^{10}–2^{14} daughter cells.

Most experiments to study survival probabilities are based on a so-called dilution assay, which briefly consists of the following steps (see also Fig. 6):

- After irradiation, a cell suspension is produced by removing the cells grown on the bottom of the culture vessel by controlled enzymatic digestion. The cell number in the suspension is counted.
- From the dose delivered, the expected fraction of surviving cells is estimated. The cell suspension is then diluted and aliquots are reseeded to new culture vessels at a density, that approximately 100 surviving cells are expected per culture vessel.
- Cells are incubated for 5–14 days typically, corresponding to 10–14 cycle times.
- The number of colonies with more than 50 cells is determined; the fraction of surviving cells is then calculated by normalization to the number of cells originally seeded in the flask.

When plotting the fraction of surviving cells as a function of dose, a typically shoulder shaped dose response curve is observed for most cell types (Fig. 6). This type of dose response curve is usually described by a linear-quadratic approach:

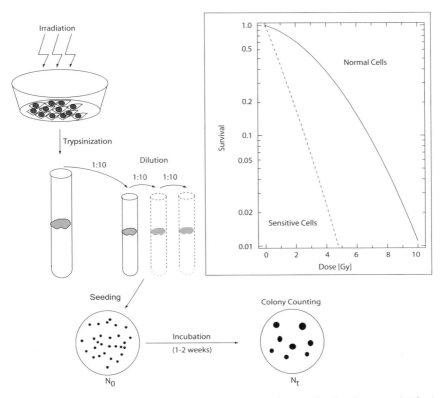

Fig. 6. Dilution assay for measuring cell survival after irradiation (for details see text). The *insert* shows typical dose response curves observed after irradiation with photon radiation of normal, repair-proficient cells (*full line*) and repair-deficient, sensitive cells (*dashed line*)

$$S(D) = e^{-(\alpha D + \beta D^2)}$$

The shoulder shape indicates that the efficiency of radiation increases with dose, which can be attributed to the more complex and thus less reparable damage induced at higher doses. This higher complexity can be explained by the interaction of damage produced in close vicinity: the probability to induce multiple 'sublethal' damage in close vicinity increases with increasing dose. This view is further supported by the results obtained with cell types containing genetic deficiencies, e.g., in DNA double-strand break repair. The loss of repair capacity is reflected in a higher overall sensitivity of the cells on the one hand, but also in a different shape of the survival dose response curve on the other hand: the repair deficient cells show a more or less straight line dose effect response [36]. The reduction of the shoulder can be attributed to the loss of repair capacity; due to this loss, even non-

complex damage cannot be repaired, so that already low doses exhibit a comparably high efficiency of cell killing.

Considering the repair of DNA damage, it might be useful to recall some numbers mentioned already in Sect. 2.2. An approximate number of 35 DSB is produced in a typical mammalian cell by 1 Gy of photon radiation (estimated from
data presented in [37]), and the surviving fraction is in the order of 50% at this
dose level. Assuming that damage is randomly distributed among individual cells,
from Poissonian statistics it can be derived that a surviving fraction of 50% is
equivalent to less than one lethal event per cell per Gy on average. That means that
at least in mammalian cells only 1 out of 35 DSB can be considered to be lethal.
One major cause of this reduction is the repair of DNA damage. However, it also
has to be taken into account that not all regions of the DNA molecule seem to be
equally important with respect to the information coded on the particular regions.
Furthermore, one important result the human genome project has shown, only a
few percent of the DNA are coding for proteins [38, 39].

2.3.4
Modification of Radiosensitivity

Several environmental factors such as, e.g., temperature, pH, or oxygen supply
have significant impact on radiosensitivity. Figure 7 illustrates the effect of oxygen
by comparing survival curves obtained under normal oxygen supply and under

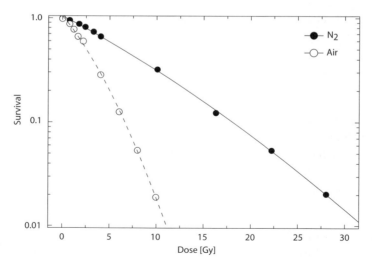

Fig. 7. Influence of oxygen supply on radiosensitivity of Chinese hamster cells. (Redrawn from
[40])

hypoxic conditions, where culture vessels were gassed with nitrogen before and during irradiation. A huge difference in sensitivity by a factor of about 3 is observed, expressed as the ratio of doses under normoxic and hypoxic conditions to achieve the same biological effect. This ratio is called 'oxygen enhancement ratio' (OER). Typically, the OER is in the order of 2.5...3.5 for most cellular systems.

When looking at the concentration dependence of the OER, it has been shown that a steep decline of sensitivity is observed only when reducing the oxygen concentration below the 1% level. In contrast, increasing the oxygen concentration from normoxic conditions (20%) to 100% O_2 has no detectable effect on radiosensitivity.

Several other substances show either radiosensitizing or radioprotective effects and are thus of interest for, e.g., radiotherapeutic applications. On the one hand radiosensitizers can be used to improve tumor control [41, 42]; on the other hand normal tissue could be spared by the use of appropriate radioprotectors. However, many radioprotectors show severe side effects themselves, making their routine use questionable [43]; only a few substances are used in clinical studies up to now [44].

2.3.5
Molecular Techniques

Several other techniques have been developed to study the radiation effects in more detail on a molecular level [45, 46]. For example, DNA sequences are studied in order to detect mutations induced by irradiation [47, 48]. Several types of mutations can be distinguished, comprising, e.g., point mutations where only a single base pair is affected up to deletions, where a whole region is cut out of the sequence.

Furthermore, consequences of the DNA damage to the complex control processes within a cell can be studied by measuring changes of the induction of protein expression on different levels [49, 50]. As a first part of the chain, the amount of messenger RNA can be detected.

Recently developed microarray techniques allow one to study several hundred types of mRNA simultaneously. Variations of mRNA may lead to a corresponding variation of the protein amount; however, this is not necessarily so, and therefore protein levels are usually studied in parallel. In addition, the availability of mutant cell lines with defined mutation patterns allows one to investigate the role of individual proteins in more detail [36, 51, 52].

2.4
Radiation Response on the Tissue Level

Up to now, radiation damage to single individual cells has been considered. An implicit assumption for many experiments is that all cells of a population can be described by the same average radiosensitivity parameters. This is not at all true when dealing with more complex tissues, which are characterized by mixtures of different cell types and a corresponding mixture of sensitivities. The particular role of stem cells has already been described; they represent the basic important unit with respect to recovery of the tissue from externally induced injury. However, normal tissue response is often described in terms of more complex 'functional subunits' [53]. They represent autonomous entities that are assumed to be able to regenerate from a single surviving cell, and different types of structural tissue organization are discussed. For example, the central nervous system represents an example of a so-called 'serial' organ, where the individual subunits are connected like the links of a chain. The damage to a single link can then already lead to serious disturbance of the organ function. At the other extreme, in parallel organized tissues the failure of a single subunit can be compensated by the function of the remaining subunits, and only when a significant fraction of subunits is inactivated will organ function be perturbed.

Although the functional subunits have not yet been biologically identified in all tissues, many of the characteristic radiation effects on tissues can be explained by the above-mentioned concept, assuming that a tissue can tolerate the loss of functional subunits up to a certain limit without detectable response. Irradiation leads to a reduction of functional subunits, and this limit thus corresponds to a threshold dose below which no effect is visible. With increasing dose, the probability of observing a certain level of response will increase, until for still higher doses, the fraction of functional subunits becomes so low that the probability of observing the effect will increase to 100%. The dose response is thus characterized by typically sigmoid curves as shown schematically in Fig. 8. However, it is not only direct damage to the functional subunits of a specific organ, but also damage to the blood vessels and the vascular structure, which might ultimately lead to a visible effect due to the corresponding disturbance of the supply with oxygen and/or nutritional factors [54].

An important parameter directly related to the kinetics of cell proliferation is the latency period, i.e., the time interval between irradiation and the occurrence of a clinically observable tissue damage. In general, the latency period is shorter for tissues with high proliferative rate like skin and mucosa, because damage leading to the loss of the proliferative capacity is expressed early ('early responding tissues'). In contrast, tissues with slowly proliferating cells require longer times to express the damage ('late responding tissues'). Furthermore, in general tissues with

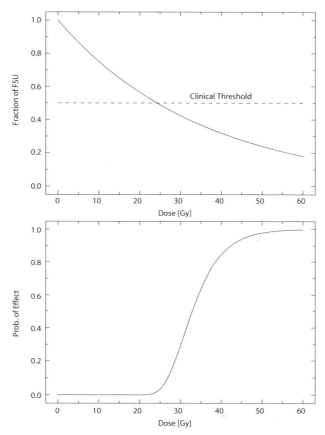

Fig. 8. Schematic representation of the dose dependence of the fraction of functional tissue subunits (FSU). A response can be clinically detected only, if the fraction is reduced below a certain threshold (*top*). This leads to the typical sigmoidal response curves, when the probability to induce a given effect is plotted as a function of dose (*bottom*)

a high proliferation rate are more sensitive to radiation compared to the slowly proliferating tissues.

It has been shown that the response of tumor tissue to irradiation can also be well described in terms of the stem cell concept [55]. With respect to the heterogeneity of the cell population, hypoxic cell fractions play a particularly important role in tumor tissues. Since many tumor types are characterized by an insufficient vascularization compared to normal tissues, they can contain substantial fractions of hypoxic cells. Due to the significantly increased resistance of hypoxic as compared to oxic cells, this cell fraction ultimately determines the probability of tumor cure.

Studies of tissue effects are usually based on animal (in vivo) experiments. Because of ethic reasons, and the considerable efforts to perform animal experiments, several experimental in vitro systems have been developed to mimic typical tissue effects as closely as possible. These comprise comparably simple systems based on a three-dimensional growth pattern distinct from the conventionally used monolayers of cells [56–58] up to quite complex systems with coculture of different cell types or utilizing the differentiation capacity of cells in vitro, leading to highly structured and compartmentalized 3D-cell systems [59].

3
Physical Characteristics of Ion Beams

Both photons and ions have in common that they induce biological damage via emission of secondary electrons when penetrating matter. However, ion beams show a quite distinct spatial distribution of energy deposition compared to photons on a microscopic as well as on a macroscopic scale. These differences are responsible also for their different biological action and will thus be described briefly here.

3.1
Microscopic Features

Photons deposit their energy mostly via photo- and Compton processes, producing secondary electrons that – if their energy is sufficient – can give rise to further ionization events and thus produce tertiary and higher order generation electrons. Since the energy deposition of an individual photon within a cellular volume is comparably small, many photons contribute to the total energy deposition when doses are in the order of 1 Gy or higher. This leads to a random spatial distribution of energy deposition events even throughout volumes of the size of typical cellular dimensions. Thus, the expectation value for an energy deposition in any small subvolume of the nucleus is almost constant throughout the nucleus.

The situation is entirely different for ion irradiation, as shown in Fig. 9. Ions also deposit their energy by emission of secondary electrons from the target atoms; however, the spatial distribution of these electrons leads to an extremely localized energy deposition along the trajectory of the primary ion. This is a consequence of the fact that the emission of electrons is peaked in a forward direction and that the overwhelming fraction of δ-electrons emitted at large angles has comparably low energies and thus short ranges [61]. Only a few electrons are produced at higher energies, which can transport energy to more distant locations from the trajectory.

The maximum range of the highest energetic electrons increases with the specific energy of the primary particle. The radial distribution of energy deposition within a particle track can be described in terms of the average energy deposition

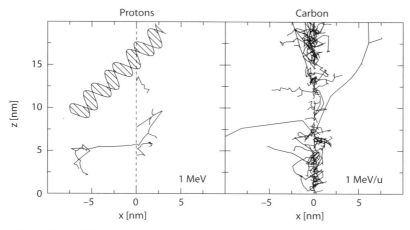

Fig. 9. Simulation of δ-electron emission by low energetic protons (*left*) and carbon ions (*right*) when penetrating tissue; tissue is assumed to be water equivalent. The *dashed line* at x= 0 represents the trajectory of the primary ions, the *full lines* represent the paths of individual secondary electrons. The size of a DNA-helix is shown for comparison in the *left panel*. (Courtesy M. Krämer; for details see [60])

d(r) as a function of the distance r from the track center, the so called radial dose profile, as shown in Fig. 10.

A characteristic feature of the radial dose profile is the $1/r^2$-dependence of the local dose. Only for very small distances is there a leveling off, and for large distances the maximum range electrons limit the distribution. The maximum range can be described by a power law:

$$R_{max} = 0.05 \cdot E^{1.7}$$

where R_{max} is measured in μm and E is the specific energy of the ion in MeV/u [62].

In the longitudinal direction particles are usually characterized by the linear energy transfer (LET), representing the energy deposited per unit track length. Although not being exactly identical, for the purposes of the present paper LET can be thought to be equivalent to the stopping power dE/dx.

It is often useful to plot biological effects not only as a function of dose, but as a function of the particle fluence F, which are related by

$$D = 1.602 \cdot 10^{-9} \cdot LET \cdot F \cdot \frac{1}{\varrho}$$

where the dose D is given in Gy, the linear energy transfer LET in keV/μm, the fluence F in $1/cm^2$, and the density ϱ in g/cm^3.

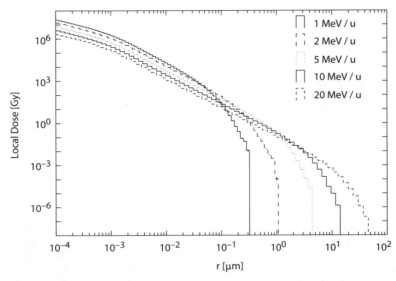

Fig. 10. Local dose deposited in carbon ion tracks at different specific energies as indicated in the legend (Courtesy M. Krämer)

In order to allow a more intuitive understanding of the particular biological effects of ion radiation, Fig. 11 compares the microscopic distribution of energy deposition of photons and ions for the typical dimensions of a cell nucleus. All four panels represent the same *average* dose, but due to the decreasing LET with increasing energy of the ion, the fluence increases as well at a fixed dose according to Eq. (3).

For photons the distribution is flat, reflecting the homogenous distribution of the expectation value of energy deposition. In contrast, low energetic ions exhibit the other extreme: in very small areas around the particle tracks extremely high local doses up to 10^6 Gy are deposited, which is compensated by large areas between the individual tracks where no energy is deposited at all. With increasing specific energy of the ions, the LET decreases and the track radius increases. Therefore, increasing overlap of contributions from different tracks is expected, which in combination with the increasing fluence leads to decreasing heterogeneity of the microscopic dose distribution. If the energy were to be further increased, in the asymptotic behavior the distribution would then resemble the flat distribution as described for photon irradiation.

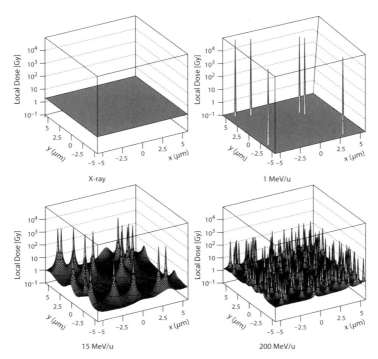

Fig. 11. Microscopic dose deposition of photos (upper left) and carbon ions at different specific energies. The size of the box corresponds to the typical size of mammalian cell nuclei. All distributions are normalized to the same *average* dose deposition of 2 Gy.

3.2
Macroscopic Features

Besides these distinct microscopic features, particle radiation also differs considerably from photon radiation with respect to the macroscopic distribution, i.e., the depth dose distribution. The typical shape is caused by the velocity-dependent stopping power, as described by the Bethe-Bloch-formula [63, 64]:

$$\frac{dE}{dx} \propto \frac{1}{\beta^2} \cdot Z_{eff}^2$$

where $\beta = v/c$ and Z_{eff} represents the effective charge of the ion. Figure 12 shows a compilation of dE/dx-values for different ion species in the energy range 1–1000 MeV/u. The energy loss is comparably low at high energies and thus velocities. With increasing depth in tissue, the velocity is reduced and thus the energy loss increases up to a maximum just before the maximum range of the ion ('Bragg-

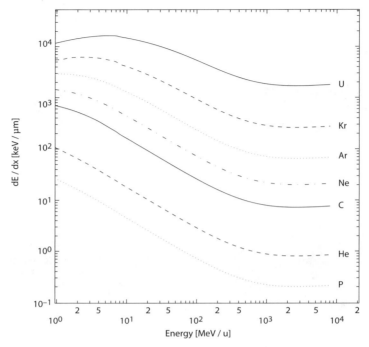

Fig. 12. Energy loss of different particle types as a function of specific particle energy. (Based on data presented in [65])

peak'). At very low energies, the originally high charge of the ion is decreased by electron capture, so that according to the Z^2-dependence dE/dx drastically drops towards the end of the particle track.

Figure 13 compares the typical depth dose profiles of carbon ion beams at different energies with the depth dose distribution of photons produced using a 20-MeV electron linac. The small tail of dose deposition beyond the Bragg-peaks is due to the nuclear interaction of the projectiles with the target atoms, leading to fragmentation of the primary ions into lighter particles. These lighter particles have a larger range compared to the primary projectiles and thus can deposit energy at greater depth.

Within the framework of therapeutical applications of ion beams, this depth dose profile is often called 'inverted', because it shows low energy deposition at the entrance to tissue and high dose deposition in depth, in contrast to photon radiation, where in general the dose deposited in depth is lower than the dose deposited in the entrance region. The only exception to this general rule is the build-up effect for very high energetic photon beams, which can be explained by the forward

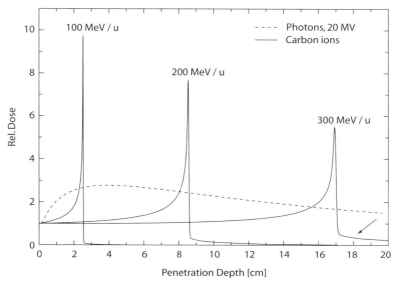

Fig. 13. Comparison of depth dose profiles for photon and carbon ion irradiation

peaked emission of high energetic secondary electrons; the secondary electron equilibrium is only reached after a few millimeter or centimeter penetration depth.

As will be explained later in more detail, the specific depth dose curve makes ion beams extremely suitable for radiotherapeutic applications to deep seated tumors because of the higher ratio of dose deposited in depth (and thus in the tumors) as compared to the dose deposited in the surrounding normal tissue. Besides the depth dose profile, further effects like, e.g., extremely small lateral and range straggling allow a very precise conformation of the dose to the tumor.

3.3
Neutron Beams

Neutrons show similar effects with respect to their biological action as charged particle beams, since their biological action is caused by the secondary charged recoil particles produced by nuclear interactions of the neutrons when penetrating matter. The spectrum of secondary particles consists mainly of low energetic protons and – depending on the initial neutron energy – a small fraction of heavier charged recoils from helium to oxygen [66, 67]. Therefore, biological effects observed with low energy, light- to intermediate charged particle beams, can explain many radiobiological aspects of neutron irradiation. However, the depth dose distribution of neutron beams is similar to that of photons, so that the dose conformation achieved with neutrons is considerably worse compared to the conformation achieved with ion beams.

4
Biological Effects of Ion Irradiation

This section first reviews the basic systematics of ion irradiation effects on single cells. The dependence of the biological effectiveness on dose, ion type, LET, and the influence of the cell type will be described. All effects will be discussed with respect to the particular physical characteristics of ion beams, i.e., track structure. The discussion includes description of the influence of environmental factors such as, e.g., oxygen supply. In addition, some recent results regarding the direct visualization of the extremely localized biological action of charged particles will be presented. The last paragraph then summarizes aspects of normal and tumor tissue response to charged particle irradiation.

4.1
The Relative Biological Effectiveness (RBE): Definition

The specific biological effects of charged particle radiation were recognized as early as 1935 [68]. However, systematic studies have been performed only after accelerators became an important tool for nuclear physics studies and could then be used also as radiation source for radiobiological applications [69–71]. A typical survival curve as obtained after irradiation of Chinese hamster cells with low energetic carbon ions is shown in Fig. 14 in comparison with the dose response curve after photon irradiation. Two essential differences are clearly visible:

- Cells respond significantly more sensitive to carbon irradiation compared to photon irradiation, i.e., carbon ions show an enhanced biological effectiveness compared to photons.
- The shape of the survival curves differs considerably. Whereas for photon irradiation the typically shouldered shape is observed, low energy carbon irradiation leads to an approximately linear dose response curve.

The enhanced effectiveness of ion beams is described in terms of the *Relative Biological Effectiveness* (RBE), defined as the ratio of absorbed doses, which have to be applied for the two radiation types to achieve the same biological effect:

$$RBE = \frac{D_\gamma}{D_I}\bigg|_{Isoeffect}$$

In most cases, 200 kV or 250 kV photon radiation is used as reference radiation. The respective biological effect level has to be defined, because the different shapes

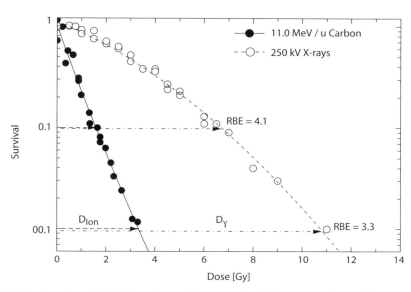

Fig. 14. Explanation of RBE. Experimental data: survival of CHO-K1 Chinese hamster cells (Experimental data from W. Kraft-Weyrather)

of survival curves imply a variation of RBE with the dose or effect level (see below). As discussed in Sect. 2.3.3, shouldered shaped dose response curves indicate the repair capacity of cells; the loss of the shoulder shape for low energetic carbon ions is thus a first hint that increased biological effectiveness is related to decreased repair capacity. In principle, this can be explained by the extremely high local doses inside the individual particle tracks as illustrated in Figs. 9 and 11. The large number of secondary electrons traversing the DNA is expected to lead to correlated damage in close vicinity, which are more difficult to repair.

4.2
Systematics of RBE

4.2.1
Dose Dependence

Due to the different shape, the survival curves cannot be simply transformed into each other by a common scaling factor applied to the dose values. Since the dose-response curve for low energy carbon irradiation can be described by a linear term:

$$S = e^{-(a_I D)}$$

the limiting RBE for D→0 is given by the ratio of the linear coefficients a_I and a_γ:

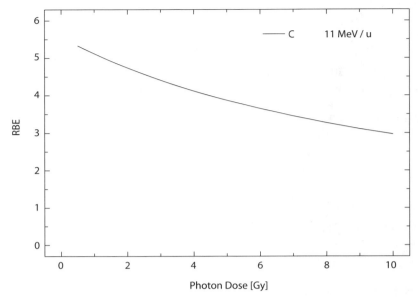

Fig. 15. Dose dependence of RBE for the data shown in Fig. 14

$$RBE_{max} = \frac{\alpha_I}{\alpha_\gamma}$$

With increasing dose, the RBE decreases, reaching an asymptotic value close to one for very high doses. Figure 15 shows the dose dependence of RBE for the example shown in Fig. 14. It should be pointed out here that although RBE varies with dose, this does not imply actually a variation of the effectiveness of the charged particle radiation with dose. Instead, it merely reflects the increase of the effectiveness of the *reference* radiation with dose, as expressed in the nonlinearity of the dose response curve in this case.

4.2.2
Energy/LET Dependence

The increased RBE is not unique for all different kinds of charged particle radiation. Instead, it strongly depends on the particular physical characteristics of the ion beam as determined, e.g., by the energy and LET of the particles under consideration. This is demonstrated in Fig. 16, where survival curves of Chinese hamster cells after irradiation with 11 MeV/u, 77 MeV/u, and 267 MeV/u carbon ions are compared. Obviously RBE decreases with increasing particle energy. Since energy

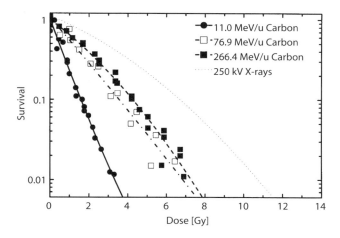

Fig. 16. Variation of biological effectiveness of carbon ions with specific energy. (Experimental data redrawn from [72])

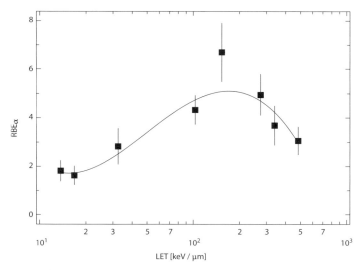

Fig. 17. Variation of RBE with linear energy transfer (LET). RBE$_\alpha$ represents the limiting RBE at very low doses, defined by the ratio α_I/α_X of the linear coefficients (Redrawn from [72])

is correlated with LET (see Fig. 12), this transforms into a rise of RBE with LET, as demonstrated in Fig. 17. However, RBE increases only up to a certain maximum at an LET of approximately 250 keV/μm, and beyond the maximum it decreases despite the further increase of LET.

Such systematics can be explained based on the known systematics of charged particle track structure as illustrated in Fig. 10. The increase of RBE from low to intermediate LET-values can be attributed to increasing energy concentration within the particle tracks; it is a consequence of the increasing LET itself and the simultaneous decreasing track diameter with decreasing specific energy. The correspondingly higher ionization density leads to more complex damage, which is more difficult to repair and thus results in the increased effectiveness.

4.2.3
Particle Dependence

Although often assumed to represent the essential parameter, LET is not suitable to define uniquely the increased biological effectiveness of charged particles (see Fig. 18). For example, the action of protons has been studied in detail and compared to the effectiveness of α-particles at the same LET [73, 74]. Protons show a significantly higher effectiveness compared to alpha particles in the LET region 20–30 keV/μm, where protons already have reached maximal RBE values; in contrast, this maximum is shifted for alpha particles to about 100 keV/μm. The same systematics extends to even heavier particles: for neon ions, the maximal RBE values are found for LET values of approximately 250 keV/μm [75].

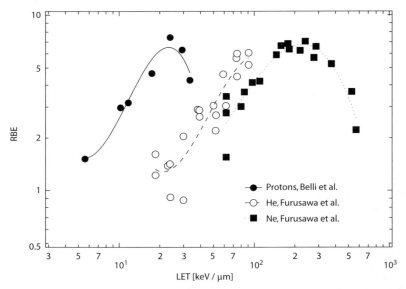

Fig. 18. Dependence of RBE on particle type. (Data for protons: redrawn from [74]; data for helium and neon ions according to [75])

These differences can be explained by similar arguments as already mentioned in the last section. When comparing different particles at the same LET, it has to be taken into account that according to Eq. (4) and Fig. 12 the particles have different specific energies. Thus, the corresponding track radii are different, and the same amount of energy is distributed in a larger track volume in the case of the higher energetic particle. Since the heavier particle has the higher energy for a given LET, it thus shows the lower energy density and thus the lower RBE.

4.2.4
Cell Type Dependence

One important aspect of RBE is that it is not solely determined by the physical properties of the ion beams; the biological characteristics of the particular cell type under consideration also plays a key role. This is demonstrated in Fig. 19 where the RBE as a function of LET is compared for three different cell types. All three cell types have a similar origin; they are derived from Chinese hamsters, but are from different organs (V79: lung fibroblasts; CHO: ovary cells) and one of them carries a genetic deficiency (XRS: genetic variant of CHO). V79 and CHO cells are genetically wild type and are repair proficient. In contrast, XRS cells are characterized

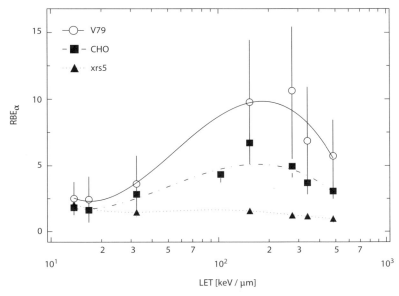

Fig. 19. Dependence of RBE on the biological system. V79-, CHO-cells: repair-proficient normal cells; XRS-cells: repair-deficient, sensitive cells (redrawn from [72])

by a mutation of a gene coding for a protein involved in the recognition of DNA double strand breaks [76, 77]. As a consequence, the repair capacity of DNA double strand breaks is diminished in XRS cells as compared to V79 and CHO cells.

These genetic differences lead to correspondingly different radiosensitivities after X-irradiation: whereas XRS cells are most sensitive and show an almost linear dose response, V79-cells are most resistant, and their dose response is characterized by a pronounced shoulder. In contrast, the dose response curves for high-LET radiation become very similar, and only minor differences are observed between XRS and V79 cells [72]. However, due to the differences observed for the reference (photon) radiation, RBE values are different for three cell lines, despite their similarity of response to high LET radiation. Whereas V79 and CHO cells show a significantly enhanced RBE, for XRS cells almost no increase of RBE is found, although the physical characteristics of the charged particles is exactly the same as for the experiments with V79 and CHO cells. However, the XRS cells share the tendency of decreasing RBE for LET values above 200–300 keV/μm with the other cell types.

The explanation for the missing increase of RBE is not as obvious, because at first glance one should expect that the same arguments concerning the increased ionization density within the tracks should hold true for all cell types. However, it has to be kept in mind that an essential part of the explanation was the expected increased complexity of the damage induced by high ionization densities, which are more difficult to repair. If, however, a cell shows a deficiency related to damage repair, increased complexity might not transform into increased effectiveness, since already non-complex damage cannot be repaired.

4.3
Very Heavy Particles

Up to now we have focused on the biological action of light to intermediate heavy charged particles in the range from proton to neon ions. As already indicated in Fig. 17, after reaching a maximum at approximately 200 keV/μm, RBE does not further increase but instead decreases towards higher LET values. Such systematics continues for heavier particles. Figure 20 compiles RBE values over the whole spectrum of particles from He up to uranium; for the heaviest particles RBE values as low as 0.05 are observed. These low values are in particular puzzling with respect to the arguments concerning ionization density mentioned above: they seem to be totally inconsistent with the extremely high energy density within, e.g., uranium particle tracks.

For the explanation of these low RBE values it has to be taken into account that the same average dose corresponds to different particle fluences according to Eq. (3). For example, a dose of 10 Gy corresponds to 4×10^7 particles/cm^2 for 11 MeV/u carbon ions (LET: 153 keV/μm), but only 4×10^5 particles/cm^2 for

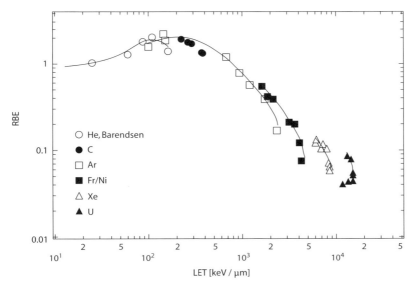

Fig. 20. RBE as a function of LET for particle species from helium up to uranium. (redrawn from [78])

5 MeV/u uranium ions (LET: 15,000 keV/μm). As a consequence, different average numbers of particle traversals per nucleus are expected.

The traversal probability is solely determined by geometrical factors, i.e., the size of the critical target which is assumed to be the cell nucleus. The average number of hits for a given fluence is then:

$$\bar{n} = A.F$$

where A is the nuclear area and F represents the fluence.

Since the nuclear size of mammalian cells is in the order of 100 μm², fluences of approximately 10^6 particles/cm² have to be applied in order to achieve an average of 1 particle traversal per cell nucleus.

In order to investigate the effectiveness of individual particle traversals, the biological effect has to be compared on the basis of the same particle fluence instead of the same average dose. Figure 21 compares the difference of both approaches for low energetic carbon and uranium ions. Whereas carbon ions are considerable more effective with respect to the average dose deposition, uranium ions are more effective when compared on a per particle basis.

The low RBE values for uranium ions reflects the fact that the dose deposited by a single particle traversal is much higher than necessary to kill the cell, leading to saturation or 'overkill' effects. On the other hand, due to the comparably low flu-

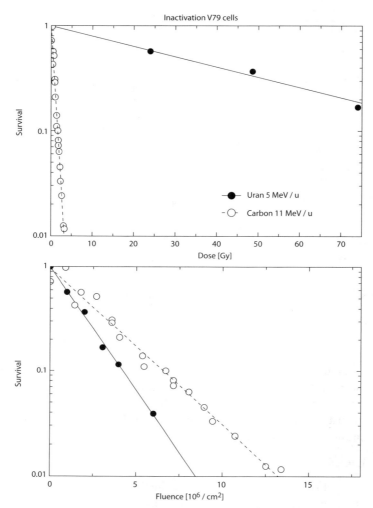

Fig. 21. Comparison of dose-response curves (*top*) and fluence-response curves (*bottom*). Both panels are based on the same data; fluences are converted to doses according to Eq. (3)

ences, a certain fraction of cells will not be traversed by a particle at all because of the stochastic distribution of traversals. Consequently, these cells are not damaged and thus will survive. A combination of both effects leads to the apparently low effectiveness of cell killing as expressed by their low RBE.

Since the probability of particle traversals is determined only by the fluence, but is independent of the particular LET, all particles above a certain LET should exhibit similar fluence effect curves which is consistent with the experimental results.

4.4
Inactivation Cross Section

The great importance of fluence has led to the introduction of a fluence related parameter to describe radiation effects of charged particles. Since empirically it is
found that dose- or fluence effect curves are linear for heavier particles at low energies, a single parameter is sufficient to represent the survival curve: the inactivation cross section σ. It is defined as the slope of the fluence effect curve:

$$S = e^{-\sigma_I F}$$

(In cases of lighter particles and/or higher energies, where curves are still shouldered, the final slope of the response curve is usually taken as measure for the cross
section.)

Fluence and dose response curves are interrelated by a transformation of LET
in dose, so that using only the linear term of Eq. (1) we get:

$$\sigma_I F = \alpha D = \alpha \cdot 1.602 \cdot 10^{-9} \cdot LET \cdot F \cdot \frac{1}{\varrho}$$

4.5
Systematics of Inactivation Cross Sections

Comparison of inactivation cross sections allows a clear overview over the systematics of biological action of charged particles. Figure 22 displays σ_I as a function of
LET over the whole range of particles from helium up to uranium.

The data show a quite complex behavior of σ as function of Z and LET and resemble the structures of the RBE(LET)-dependence shown in Fig. 20. For comparison, the dashed line represents the expected dependence, if RBE were constant for
all particles. This corresponds to a constant value of α, and according to Eq. (10)
the inactivation cross sections should then simply increase proportional to LET.
The data lie above this line in the LET range from 50 to 500 keV/μm. For higher
LET values and very heavy ions, the data exhibit a pronounced hook structure, but
lie significantly below the dashed line and thus reflect the extremely low RBE values shown in Fig. 20.

To facilitate the explanation of these structures, the same data rare redrawn in
Fig. 23 now as a function of the specific particle energy. It is clearly visible, that for
the extrapolation to low energies inactivation cross sections show a convergent behavior to a common value of approximately 40 μm². With increasing energy, the
curves successively diverge. For very heavy ions like, e.g., uranium, an increase of
σ is observed. For krypton ions this rise is already significantly reduced, and for
light particles like carbon the dependence is inverted, i.e., decreasing σ are observed with increasing energy.

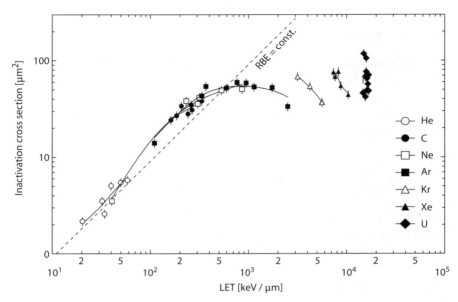

Fig. 22. Inactivation cross section for V79 Chinese hamster cells as a function of LET. (Data redrawn from [78])

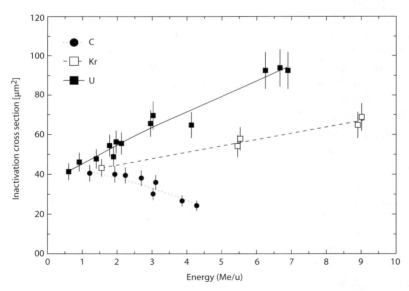

Fig. 23. Inactivation cross section for V79 Chinese hamster cells as a function of specific particle energy. (Data redrawn from [78])

Again, track structure is the key to understand the complex dependencies seen in Figs. 22 and 23. For E→0, $\sigma(E)$ is approximately the same for all ions in the region C-U. The reason for this similarity is the small track radius for low energies and the comparatively high LET. In this case, energy can be deposited in the cell (nucleus) only with direct traversals, and the probability of traversals is solely determined by the geometric cross sectional area of the nucleus. If LET is sufficiently high, each individual traversal will lead to inactivation with high probability, and inactivation probability depends only on the hit probability, but not on the particular LET. Therefore, inactivation cross sections are similar for different particles.

With increasing energy, the track radius increases as well according to Eq. (2). Then even particles with an impact parameter $d>r_{nucleus}$ can deposit energy in the nucleus. For very heavy ions, energy deposition from those indirect hits might still be sufficiently high to kill the cell, because the local dose at the boarder of the track is still considerable. Thus, in this case inactivation cross-sections rise with particle energy. Although the track width is the same also in the case of carbon ions, the local dose deposited at a certain distance from the track center is considerably lower compared to the case of uranium ions due to the lower LET. In this case, the energy deposition is too small to lead to inactivation for large impact parameters $d>r_{nucleus}$. In addition, according to Fig. 12, LET decreases with increasing specific energy, so that even a direct hit may not be sufficient to kill a cell. Therefore inactivation cross-sections decrease with increasing energy in the case of lighter ions like, e.g., carbon. For intermediate heavy particles both of the effects described above contribute, so that decrease of LET more or less compensates for the increasing track diameter.

Using Fig. 12, the systematics described above can be converted to the systematics of σ as a function of LET. The characteristic hook structure thus is a consequence of the variation of track diameter with particle energy. The dominance of the physical aspects of track structure at very high LET is further supported by the similarity of the $\sigma(LET)$-dependencies for different biological systems and endpoints as summarized, e.g., by Kraft [78]. In general, they all exhibit a pronounced hook structure for ions heavier than argon as demonstrated also in Fig. 22.

4.6
Oxygen Effect

For low-LET radiation a pronounced dependence of radiosensitivity on the oxygen supply is observed. This effect is largely reduced for high-LET radiation, and nearly vanishes for LET>1000 keV/μm (see Fig. 24). This feature of high-LET radiation is expected to be a remarkable advantage in tumor therapy. The resistance of hypoxic cells, even if only contributing as a very small fraction to the total tumor mass, can limit the probability of tumor cure in the case of conventional radiation. In contrast, for high-LET radiation, according to the results described above, hypoxic

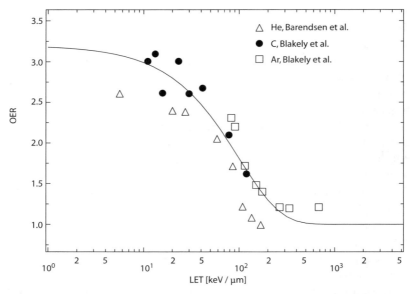

Fig. 24. Dependence of oxygen enhancement ratio on LET for different particles. (Data redrawn from [79, 80])

cells exhibit a similar sensitivity as normoxic cells and should thus allow a significant increase of tumor control in the case of tumors with hypoxic cell fractions.

4.7
The Role of Increased Ionization Density

The explanation of the systematics of RBE and σ as a function of particle energy and LET were largely based on arguments concerning ionization density. Increased ionization density was assumed to lead to more complex and thus less reparable damage. Although plausible, these arguments need of course further confirmation from experimental data. This section summarizes experiments showing more directly the influence of the increased ionization density of charged particle radiation on the complexity and reparability of damage.

4.7.1
Double Strand Break Induction and Rejoining

Intuitively one would expect the increased ionization density to lead to a higher yield of severe damage, e.g., DSB. Surprisingly such a higher rate of DSB production has not been found; RBE values for DSB induction are very close to one even in LET-regions where the RBE for survival is significantly enhanced [81]. Obviously, there is no simple correlation between DSB induction and lethality. However, it has to be kept in mind that the measurements of DSB induction based on gel electrophoretic methods give average yields, but no conclusions can be drawn on the spatial distribution of DSBs within the individual cells. For example, when considering a given number of DSBs, their biological effect might well depend on whether they are all produced as clusters in a small subvolume of the nucleus (as would be expected for particle tracks) or homogeneously distributed throughout the nucleus. Clusters of DSB constitute a more complex type of damage which is more difficult to repair. This is consistent with results from studies of the rejoining of DNA fragments.

These studies indicate that, although the initial yield of DSB is similar for different radiation types, the residual damage after high-LET radiation is usually higher due to the reduced reparability of the complex damage induced. Residual damage is thus thought to be the more appropriate measure of lethality than the primary induced damage.

Figure 25 gives typical examples of the rejoining kinetics after X-irradiation and irradiation with low energetic Ni ions. Both the time constant as well as the fraction of residual damage are significantly increased for the low energy nickel irradiation. This finding is fully compatible with the changing slope of the survival

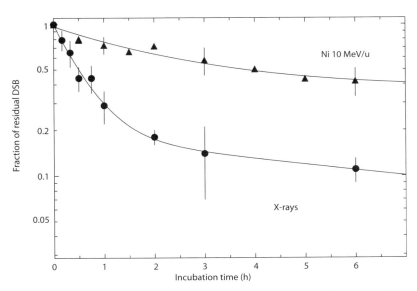

Fig. 25. Influence of radiation quality on the rejoining of double strand breaks in CHO cells. The relative fractions of DSB are normalized to the initial number of DSB observed at t=0. The dose values for photon and nickel ion irradiation were adjusted to induce approximately the same absolute number of DSB at t=0. (Courtesy G. Taucher-Scholz)

curves, the vanishing shoulder at low energy being an indicator of the more complex and thus less reparable damage.

Although the standard assay for detection of rejoining does not lead to conclusions about the fidelity of the rejoining process, more detailed investigations revealed that close vicinity of DNA damage actually leads to a significantly higher rate of misrepair processes [82].

However, it is not only the spatial vicinity of DSB which might influence the effectiveness. In particular combinations of different types of damage, e.g., DSB plus base damage or even clusters of damage containing no DSB, which are produced much more frequently, are discussed as relevant complex damage [22].

4.7.2
Chromosome Aberrations

Investigation of chromosome aberrations also revealed the influence of radiation quality on the rate of rejoining and repair of DSB. For example, the yield of exchange type aberrations, requiring rejoining of chromosome fragments induced by at least two DSB in two chromosomes, is significantly smaller for high-LET radiation than for low-LET radiation. Complementarily, the rate of chromatid breaks is considerably enhanced [31].

In contrast, after conventional X-irradiation, the damage is distributed homogenously throughout the nucleus, so that the involvement of many chromosomes is much more likely, leading also to the higher rate of exchange type aberrations between different chromosomes.

A particularly impressive consequence of the extremely localized energy deposition is shown in Fig. 26. It shows a mitotic cell irradiated with low energetic, very heavy ions. Whereas most chromosomes appear unaffected by irradiation, in one single chromosome a huge number of breaks is induced, leading to a particular type of aberration called 'disintegration' [83]. Due to the spatial separation of individual chromosome territories during mitosis, it is very likely that a single particle traversal through the nucleus only damages a single or few chromosomes, leaving the remaining chromosomes totally unaffected. Similar patterns of chromosome damage have been also found after irradiation of cells with extremely focused laser-UV-irradiation [84].

4.7.3
Fractionated Irradiation

Further evidence for the severity and complexity of high-LET induced damage comes from studies of fractionated irradiation. Here, the total irradiation dose is given in two or more fractions with a certain time interval in between. During this time interval, cells are incubated under optimal conditions, so that at least part of

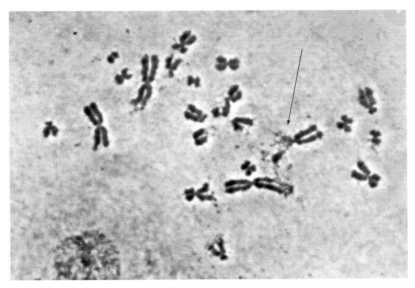

Fig. 26. Observation of chromosomal disintegration (*arrow*) after irradiation of mitotic cells with low energetic, very heavy particles. The complete destruction of parts of the chromosome can be explained by the extremely high ionization density within the particle track (Courtesy S. Ritter)

the damage can be repaired, leading to an overall decreased effect of fractionated compared to single dose exposure.

For photon irradiation these effects are generally very pronounced. In contrast, with increasing LET the fractionation effect decreases and eventually completely disappears; in some cases even an enhanced effectiveness is reported [85]. This indicates, that no repair of radation damage is possible in the interfractionation interval, supporting the view of more complex damage induced by particle radiation.

4.7.4
Direct Visualization of Localized Damage

More recently, new methods have been applied to visualize directly the localized damage induced by particle radiation. These methods are based on immunofluorescent staining of nuclear sites containing proteins involved in DNA repair. Accumulation of these proteins is thought to be an indicator of particular severe and probably irreparable damage [86, 87].

Figure 27 shows an example of such an experiment, demonstrating accumulation of the p21-protein confined to the sites of particle traversals. The beam direc-

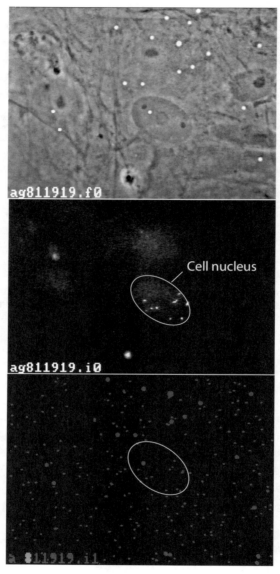

Fig. 27a–c. Visualization of the localized DNA damage by immunofluorescent methods: a phase contrast microscopic image of human fibroblast cells; b immunostaining of p21-protein after irradiation of the cells with 10.1 MeV/u Ca ions. The image is taken at exactly the same position as the phase contrast image a – the position and shape of a cell nucleus is indicated for comparison; c position of the individual particle tracks as observed in nuclear track detectors on which the cells were grown. As indicated by the outline of the cell nucleus, the immunostaining pattern shown in b reflects the track pattern in c and thus confirms, that the biological response is confined to the small regions of high ionization density within the track

tion is perpendicular to the image plane in Fig. 27, and each spot in the cell nucleus highlighted in the middle panel indicates the traversal of a single ion. This has been proven by correlation of the fluorescent spots with the image of etched ion tracks as shown in the lower panel, taken from exactly the same position of the sample [88]. Although the particular role of the p21-protein in this response to particle induced damage is not yet known, its colocalization with other proteins involved in DNA repair suggests a new role in DNA damage processing. The localized accumulation of these proteins is completely compatible with the distribution of damage expected according to the local dose distributions as shown in Fig. 11.

4.8
Bystander Effects

The term 'bystander effect' is used to describe situations where not only the primarily damaged cells respond to radiation, but also neighboring cells show a response without being directly damaged. This is particularly relevant in the case of charged particle radiation because of their stochastic properties. Diluting the average dose corresponds to a decrease of the particle fluence and thus the average number of traversals n. Therefore, in the case of n<1 it is not the dose to the individual cell which is decreased, but instead it is the fraction of cells receiving a particle hit which is reduced. Therefore, at low fluences there is a mixed population of hit and thus potentially damaged cells and unhit or unaffected cells. The bystander effect, however, can induce a response even in these unhit cells.

There is now increasing evidence for significant contributions of the bystander effect at low doses, although mechanisms are still to be elucidated. For many of these studies, special irradiation facilities (microbeam facilities) have been used, which allow one to irradiate individual cells with a predefined number of particles at high spatial resolution (<1 μm) [89–91]. On the one hand this allows one to choose a particular site within the cell (cytoplasma vs nucleus) for irradiation, on the other hand defined numbers of cells can be irradiated with defined numbers of particle traversals, thus avoiding some inherent problems related to the stochastic distribution of particle hits in conventional irradiation experiments.

Using such a facility, it has been shown for example that more damaged cells can be found than primarily irradiated [92, 93]. The comparably low cell density and the spatial distribution of damaged cells with respect to the irradiated cell suggest a soluble factor transmitted through the medium as reason for the observed bystander effect.

Other experiments revealed an increased rate of transformation or mutation compared to the expectation value based on the number of traversed cells [94, 95]. Furthermore, induction of gene expression has been studied using directed as well as stochastic particle irradiation. These experiments also indicate a more pronounced increase of the response than expected on the basis of the number of traversed cells, which is interpreted as bystander effect. Some experiments revealed, that direct cell-cell contact mediated via gap-junctions might play an important role for this type of bystander effect [94, 96, 97]. The bystander effect could also explain at least partially the hypersensitivity observed after irradiation with charged particle beams at very low doses [98–100].

However, there is not yet a clear line of evidence which could explain all the puzzling and partially contradictory results. However, heavy ion beams and in particular the microbeam facilities are ideally suited to study bystander effects.

4.9
Tissue Effects

The above-mentioned bystander effects are expected to be of importance for understanding the response of tissues to radiation injury, because cells in a tissue are usually connected by complex communication networks. In fact there are first indications that bystander effects cannot only be detected in in vitro systems, but also in in vivo-like systems such as, e.g., tissue explants [101]. Since bystander effects play a role at low doses and low fluences, they might be in particular relevant for studies of mutation and transformation related to radiation protection. For applications of ion beams in tumor therapy, doses and thus fluences are usually comparably high, so that the fraction of unhit cells is small.

Therefore, the effects of indirect injury might be masked by the comparably higher contribution of direct injury.

4.9.1
Normal Tissues

Experimental studies of heavy ion irradiation on tissues have been performed mainly for the preparation of heavy ion therapy facilities as, e.g., in Berkeley (USA), Chiba (Japan), and Darmstadt (Germany) [102–108]. Generally, it is found that the types of tissue damage induced by ion beam irradiation are similar to those found after X-irradiation, at least from the histopathological point of view [109–111]. Furthermore, the relative shape of the dose response curves is similar for both radiation qualities. However, for high-LET radiation, the curves are shifted to lower doses due to the increased RBE, i.e., the same level of effect is observed at lower doses of ion beam irradiation compared to photon irradiation. Assuming that stem cells represent the basic unit responsible for regeneration of a functional tissue subunit after radiation insult, this difference can be explained by the increased inactivation effectiveness for these stem cells. Since the time scale of development of detectable tissue damage is largely determined by the turnover rate of the particular cell system under consideration, it becomes plausible that the time scale of the development of tissue reactions is at least similar for ion irradiation and photon irradiation; some studies indicate slightly shorter latency periods for ion irradiation.

Most of the studies have been performed either using skin or intestinal irradiation on the one hand or spinal cord on the other, representing typical examples of early and late responding tissues, respectively. Similar to cellular effects, RBE values depend on several physical factors such as, e.g., particle type, dose, LET, as well as on biological factors like, e.g., tissue type and biological endpoint under consideration. With respect to application in tumor therapy, the tissue dependence of RBE is of major importance. In general, higher RBE values are found for late responding tissues than for early responding tissues [103]. Such systematics is consistent with the findings of studies using neutron radiation [112, 113]. Similar to the differences between repair proficient and repair deficient cells, the differences of RBE between early and late responding tissues can be mostly attributed to differences in the shape of the *reference* dose-effect-curve, whereas the dose response curves for *high-LET-radiation* become very similar [114].

4.9.2
Tumor Tissue

Since in general normal tissues surrounding the tumor volume are limiting the dose that can be given to the tumor, most studies of high-LET effects have been performed using normal tissue systems. Studies of experimental tumor systems have been mainly performed in the framework of heavy ion therapy at the BEVA-LAC [115–119]. In these experiments, the tumor response was studied on different biological levels including tumor cure and tumor growth delay. A particular type of experimental tumor (rhabdomyosarcoma) was chosen, which is characterized by a high potential to develop hypoxic areas and is thus suitable to exploit the potential advantages of high-LET radiation with respect to a reduced oxygen dependence of radiosensitivity. The results supported the advantage of heavier ions like neon beams for therapeutical applications in the case of tumors containing a significant hypoxic fraction. However, it has to be taken into account that neon ions would also exhibit a significantly increased RBE in the surrounding healthy tissue.

The results have stimulated further investigations on the role of hypoxic fractions in charged particle tumor therapy. Ando et al. [120] and Oya et al. [121] studied the reoxygenation of experimental mouse tumors after charged particle radiation as compared to photon radiation. Reoxygenation describes the oxygen supply of formerly hypoxic cells due to the inactivation and consequent degradation of the more sensitive oxic cell fraction during fractionated irradiation. The results obtained by Ando and Oya indicate a more rapid reoxygenation after high-LET radiation than after photon irradiation, at least for some tumor cell strains.

5
Models of Biological Action of Heavy Charged Particles

The progress in experimental studies is complemented by the development of biophysical models to describe radiation effects. Modeling plays an important role for the mechanistic understanding of radiation action as well as for practical applications in radiation protection and radiotherapy. This section will thus give a brief overview over some basic concepts developed in particular to describe the biological action of heavy charged particle beams.

5.1
Theory of Dual Radiation Action

For low LET irradiation, one of the most prominent features is the shoulder shape of the corresponding dose response curves. One possible explanation of this shape is the interaction of sublethal damage, resulting in lethal damage. Since the interaction probability increases with the density of sublethal damage, higher doses lead to a higher interaction frequency and thus higher efficiency compared to lower doses, resulting in the shoulder shape of the dose response curve.

Based on the analysis of chromosome aberrations, Neary [122] has developed one of the first models specifically based on lesion interaction. Estimates revealed that interaction should take place over distances of typically micrometers, so that the distribution of damage on a micrometer scale was thought to be of particular importance. Since the spatial distribution of damage cannot be investigated directly, the spatial distribution of energy deposition was taken as a measure reflecting the damage distribution.

Among others, the theory of dual radiation action (TDRA) is one of the best known models based on detailed considerations of damage interaction [123]. The development of the model was accompanied and stimulated by the development of detectors and experimental techniques, which allow the detection of energy deposition distribution with micrometer resolution ('microdosimetry' [124]).

However, eventually it became clear that the assumption of a micrometer interaction distance was inconsistent with more recent results from experiments using very low energetic photon radiation. These data were only compatible with considerably smaller interaction distances in the order of nanometers [125]. Since there are no experimental techniques to resolve the spatial distribution of energy deposition with such a fine resolution, the corresponding refinements of the TDRA [126, 127] had to rely on computer simulations of charged particle track structure. Surprisingly, although the model is based on very general assumptions and sophisticated mathematical treatments, quantitative comparisons of model predictions with experimental data are extremely scarce. Thus, the important role of this model has to be seen under conceptual aspects rather than practical applications.

5.2
Cluster Models

Progress in experimental techniques for investigation of molecular details of radiation damage revealed that thinking in terms of 'simple' double strand breaks as relevant lesion would be inadequate for a detailed understanding of radiation action [128]. Instead, there were hints that the molecular structure of strand breaks might vary considerably. For example, estimations revealed that the lethality of a DSB induced by high-LET radiation is significantly higher than the lethality of a DSB induced by low LET radiation [81, 129–131]. This could be due to the fact, that a DSB induced by high-LET radiation might actually consist of a combination of several DSBs or of a DSB with other damage, which however cannot be resolved with the current experimental techniques because they are produced to close together. Therefore, damage complexity is assumed to be an important key to understand high-LET effects.

Clusters of damage should then result from clusters of energy deposition, and thus several models have been developed which are particularly based on detailed investigations of cluster properties of high-LET radiation with nanometer resolution. [132–135]. At the present stage, the predictive capacity of these models is essentially restricted to the initial phase of strand break induction. Application to more complex endpoints like, e.g., cell inactivation yields only qualitative agreement [136] and would require a detailed modeling of the subsequent damage processing of the initial damage. This, however, seems not to be feasible due to the enormous complexity of the signaling pathways involved in damage processing, which also depend very much on the particular cell or tissue type under consideration.

5.3
Amorphous Track Structure Models

The above-mentioned disadvantages are bypassed in another class of models called 'amorphous track structure' models. The two key features of these models are:

- They make use of the particular features of track structure in a certain simplified form, in that they use only the information about the radial dose profile $D(r)$, but explicitly neglect the stochastic properties of secondary electron emission.
- They are based on the assumption that no principle difference between the biological actions of low- and high-LET radiation exists, because in both cases the biological effect is due to the action of the secondary electrons. Thus, it should make no difference with respect to the biological action, whether an electron is produced by a primary photon or by a charged particle; the biological effect

should be simply determined by the total amount of energy deposition in a certain, appropriately sized subregion of the cell.

Combining both aspects, the response of a cell to a charged particle traversal can thus be obtained from an appropriate convolution of the microscopic dose distribution with the low-LET dose response curve of the particular biological object under consideration.

Although based on similar assumptions as described above, the models proposed by Katz and coworkers [137–139] and by Scholz and coworkers [140–142] substantially differ with respect to the conversion of the general assumptions into a computational algorithm [143]. We will focus here on the method described by Scholz et al. [142] termed 'local effect model' (LEM).

The basic idea of the approach is schematically shown in Fig. 28. The cell nucleus, represented by the circular area, is assumed to be the critical target for cellular

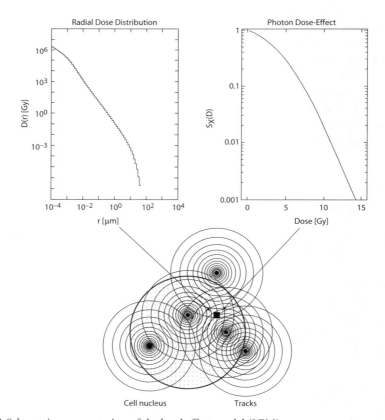

Fig. 28. Schematic representation of the local effect model (LEM)

effects, and a homogenous distribution of sensitivity throughout the nucleus is assumed as a first approximation. The cell nucleus is virtually subdivided into small compartments, and for a given set of impact parameters the average dose deposition in each compartment is determined from the corresponding radial dose profiles of individual tracks. From the local dose at a given position, the local biological effect is calculated from the effect observed for photon irradiation at the same dose. This requires an appropriate scaling according to the ratio of the small compartment volume and the total sensitive volume.

The total average number of lethal events induced by a particular pattern of particle traversals is then calculated by numerical integration:

$$\overline{N}_{lethal} = \int \frac{\ln(S_X(d))}{V} \, dV$$

where $S_X(d)$ represents the photon dose response curve, d is the local dose at a given position in the nucleus, and V is the nuclear volume. Survival is then determined according to Poisson statistics:

$$S = e^{-\overline{N}_{lethal}}$$

This simple formula has been shown to represent all characteristic features of high LET radiation as, e.g., the complex structure of inactivation cross sections as function of LET and/or particle energy, the dose dependence of RBE, and the dependence of RBE on the particle species for a given LET. Furthermore, the model is able to reproduce the dependence of RBE on the cell type under consideration.

The latter aspect is particularly important for applications of the model in the framework of treatment planning for heavy ion therapy (see below). A comparison of the model calculations of RBE with experimental data for three different cell systems is shown in Fig. 29. The model represents the cell dependent RBE: a significant rise of RBE with LET is predicted for V79 and CHO cells, whereas for XRS cells no pronounced maximum is predicted, consistent with the experimental data.

Since the physical characteristics are exactly the same for all three cases, the differences of predicted RBE values can be entirely attributed to the different reference low-LET survival curves which are used as input for the calculation (they are shown as insert in Fig. 29). Obviously the RBE is correlated with the shoulder shape of the photon dose response curve: for pronounced shoulders, significantly enhanced RBE values are predicted, whereas with decreasing shoulder width, decreasing RBE values are expected, and in the extreme case RBE should be close to one for cell systems showing purely exponential survival curves for photon radiation [141].

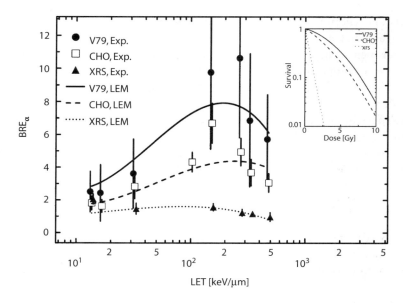

Fig. 29. Comparison of model calculations with experimental data. (Experimental data re-drawn from [72])

5.4
Influence of Target Structure

The model described above is based on the assumption that lethal damage corresponds to point-like events, where point-like is meant in a biological sense, i.e., a point corresponds to the size of the smallest relevant biological structure, assumed to be the DNA strand and thus a size of a few nanometers. The critical, sensitive target for inactivation, i.e., the DNA, is assumed to be homogenously distributed throughout the cell nucleus, and no specific substructure has been assumed.

The situation is entirely different when dealing with other biological endpoints like, e.g., induction of chromosome aberrations. Here, interaction of damage on different chromosomes plays the dominant role, and therefore details of the distribution of chromosome territories within the nucleus have to be taken into account [144].

Frequently the structure of a chromosome is simulated by a random walk polymer chain model. Given the appropriate boundary conditions, this leads to a relatively compact chromatin organization in the centromere region of a chromosome and a more sparse distribution in the outer regions [145]. As a consequence, this leads to a significant overlap and intermingling of DNA from different chromosomes.

Although earlier data seemed to support such a high degree of intermingling, more recent data indicate that chromosomes are confined in clearly separated territories within the nucleus, showing overlap only in the contact regions between two territories [146, 147]

6
Application of Charged Particle Beams in Tumor Therapy

The basic advantages of charged particle beams such as, e.g., carbon ion beams as compared to conventional photon and electron beams are the inverted depth dose profile (Bragg peak characteristics) and the increased biological effectiveness.

Proton beams only show the advantageous depth dose profile without the additional increased effectiveness (see below).

Besides the general physical and radiobiological properties as described in the previous sections, some additional specific aspects for the application to tumor therapy will be described here. These include technical aspects of, e.g., beam delivery as well as radiobiological considerations. The particular solutions developed for the heavy ion therapy pilot project at GSI will be described in more detail.

6.1
Technical Realization

A single Bragg peak as shown in Fig. 13 is too narrow for irradiation of extended tumor volumes, which typically have extensions in the order of a few centimeters. Therefore, special beam delivery techniques had to be developed in order to achieve a homogenous irradiation of extended volumes. In depth this is achieved by superposition of several Bragg peaks with different peak positions, as illustrated in Fig. 30. The position of the individual Bragg peaks is defined by the energy of the primary particles; the required energy variation can be achieved by passive as well as by active methods [148, 149].

Passive methods are based on a fixed primary beam energy and variable absorbers in the beam to degrade the energy. In contrast, active methods are based on the corresponding adjustment of the accelerator settings, so that already the primary beam energy is chosen according to the desired penetration depth.

Furthermore, the lateral contour of the treatment field has to be adapted to the contour of the tumor volume in order to reduce the dose deposition in the surrounding healthy tissue. For passive methods, collimators are used in general. In contrast, active methods make use of the electromagnetic properties of charged particle radiation, namely the possibility of magnetic deflection of the beam. Using a pencil beam of a few millimeters in width, the beam can be scanned in the x- and y-directions by dipole magnets, allowing one to cover any irregular field shape with a predefined dose.

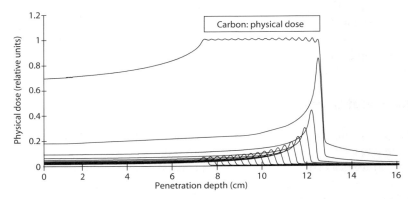

Fig. 30. Irradiation of extended volumes with homogenous dose deposition by superposition of Bragg peaks at different depths, corresponding to different beam energies. (Courtesy U. Weber)

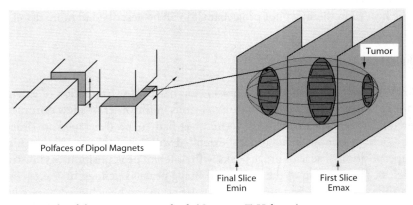

Fig. 31. Principle of the raster scan method. (Courtesy T. Haberer)

Combination of active methods in the longitudinal as well as in the lateral direction allows an optimal conformation of the dose distribution to the tumor volume. Such a system, called raster scan system (Fig. 31), has been realized for the heavy ion therapy project at GSI [150, 151]. Similarly, the proton therapy at PSI is using an active scanning method; only a corresponding movement of the patient replaces the magnetic scanning of the beam in the y-direction [152, 153].

A further advantage of carbon ion beams is the possibility to use positron emission tomography (PET) to verify the position of the treatment field in the patient quasi online during the treatment period [154]. This method is based on the small amount of radioactive, positron emitting ^{10}C and ^{11}C isotopes produced by nuclear interactions of the projectile ions with the target atoms. Both isotopes have ranges similar to that of the primary beam. Due to the half life of the radioactive isotopes most positrons are emitted from stopped ions, thus allowing one to detect the distal edge of the dose distribution with high precision.

6.2
Radiobiological Considerations

6.2.1
Optimal Ion Species

One of the key issues for application of ion beams in therapy is the choice of the appropriate optimal ion species. As discussed in Sect. 4, two effects govern the systematics of RBE: the increase of the efficiency per particle traversal with LET and the counteracting influence of the stochastic distribution of particle traversals. Therefore, at least from in-vitro studies, carbon beams are expected to represent the optimal choice for therapy. Figure 32 illustrates the rationale for this choice by comparing survival curves obtained for irradiation with different high- and low energy particle beams.

The higher energies have been chosen to represent the same penetration depth for all ion species, thus simulating damage to the tissue at the entrance channel for

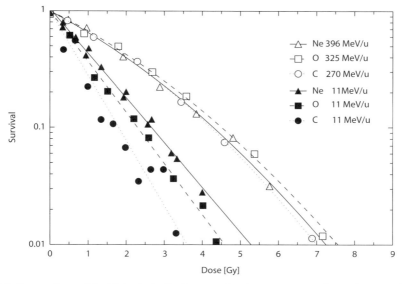

Fig. 32. Comparison of the biological effectiveness of carbon, oxygen, and neon beams at high and low energies. (Courtesy W. Kraft-Weyrather)

irradiation of a tumor at a given depth; the low energy beams represent the efficiency at the distal end of the extended Bragg peak and thus the tumor region. The effects of the high-energy beams are nearly identical. In contrast, the efficiency of the low energy particles decreases with increasing particle mass. Therefore, carbon ions are expected to show the optimal therapeutic gain in terms of the increase of RBE with penetration depth.

6.2.2
Depth Dependence of RBE

The depth dependence of RBE represents the radiobiological key feature of heavy ion beams for tumor therapy: RBE rises with increasing penetration depth, so that the RBE in the Bragg peak region in general is significantly higher compared to the RBE in the proximal part. This is illustrated in Fig. 33, which compares the physical depth dose distribution and the corresponding RBE as a function of penetration depth. The increase of RBE essentially reflects the increase of the average LET with depth.

Although the average RBE in the region of the extended Bragg peak is not as high as for a single Bragg peak, still there is a significant benefit to be expected in that the RBE in the tumor volume is higher than the RBE in front of the tumor volume. Thus, the increase of physical dose with depth is further enhanced by the rise of RBE with depth.

The situation is different in the case of proton beams. Although the principle of superposition of Bragg peaks to achieve a homogenous dose distribution over extended volumes is the same for protons and heavier ions, protons in general show significantly lower RBE values compared to carbon ions. Only for energies below 10 MeV is a significant rise of RBE observed (see, e.g., Fig. 18). Transferred to the situation of superimposed Bragg peaks, this should lead to significantly enhanced RBE values only at the very distal part of the extended Bragg peak. Therefore, in general, RBE values of 1.1 are used for correcting physical dose values to biological effective dose values in proton therapy without considering any dependence of RBE with dose, tissue type, and position in the treatment field [155].

Meanwhile there are data indicating a significantly enhanced RBE at least for in-vitro systems also in the case of proton radiation [156]; however, since the relevance of this data to the clinical situation is still under discussion, these data have so far not been considered for treatment planning in proton therapy.

This is in contrast to the case of heavier ions, where the estimation of the increased RBE values is a prerequisite for the safe application of ion beams in tumor therapy. Different approaches to take the increased efficiency into account in treatment planning for high-LET beams are reported in the literature, ranging from experimentally oriented approaches to methods based on biophysical modeling [142, 157–160].

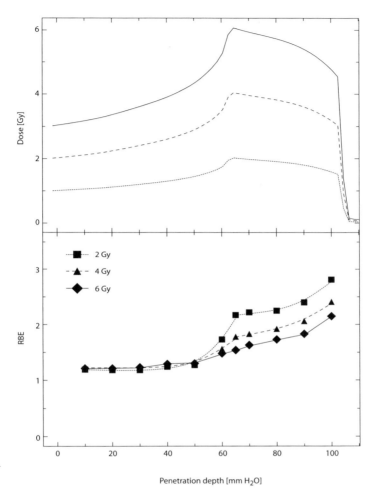

Fig. 33. Depth dependence of RBE in an extended Bragg peak region as obtained by superposition of individual Bragg peaks (see Fig. 30). (Courtesy W. Kraft-Weyrather)

One particular aspect, which has to be considered in treatment planning, is the tissue dependence of RBE. RBE of different normal tissues or tumors might be different, and therefore the therapeutic gain of charged particle beams will be ultimately defined by the particular combination of tumor type and normal tissue. According to the radiobiological considerations described in Sect. 4, slow growing, resistant tumors are expected to show the most significant benefit from high LET radiation. In addition, tumors with considerable fractions of hypoxic cells represent the second major class of tumors taking advantage of high LET radiotherapy.

6.3
Clinical Aspects

First applications of charged particle beams in tumor therapy were all performed at accelerators constructed for fundamental research in nuclear and particle physics [161, 162]. Only in the last decade the first fully hospital based treatment facilities were built in Loma Linda (USA) and Chiba (Japan). Extension of the charged particle radiotherapy program is particularly supported in Japan; several other facilities are planned there or are already under construction (for an overview see [163]).

Whereas all of these facilities use passive beam delivery techniques, at PSI (Villigen, Switzerland) and GSI (Darmstadt, Germany) the first active beam delivery systems have been successfully tested and are now used routinely for patient treatments with protons (PSI) and carbon ions (GSI). Both facilities are installed at accelerators constructed for fundamental research, with no direct clinical environment. Patient numbers are limited due to the basic research programs performed at these accelerators. At GSI, more than 100 patients with slow growing tumors in the head and neck region have been treated up to now. Since results are very promising so far [164], the installation of a dedicated clinical heavy ion therapy facility is planned at the Heidelberg University Clinics, and beginning of construction of the facility is expected for fall 2002. This facility will allow approximately 1000 patients to be treated per year and can deliver different ion beams in the range from protons up to oxygen ions. It is thus ideally suited for adaptation of the treatment according to a biological optimization, taking into account the particular relative biological effectiveness for the tumor tissue and the surrounding normal tissue, respectively.

7
Summary and Perspective

The characteristic radiobiological properties of ion beam radiation as compared to photon radiation can be explained in terms of their particular physical properties in combination with the potential of the biological object to process and repair radiation induced damage. The highly localized energy deposition of charged particles leads to complex types of damage, which are more difficult to process and repair. For cells which are able to repair the less complex damage induced by photon radiation to a large extent, the higher complexity of damage induced by charged particles is expressed as a higher biological effectiveness, i.e., lower doses are required to achieve the same effects compared to conventional radiation.

Depending on the particles atomic number and specific energy, the ionization density within the track and thus damage complexity can be varied substantially. Charged particle beams thus represent a unique tool to study the influence of damage complexity on the biological response to ionizing radiation. Despite the extensive studies performed in the last decades, the molecular structure of damage leading, e.g., to cell lethality is not yet known. However, there is probably no unique answer to this problem, because the consequence of a certain primary damage critically depends on the potential to process and repair damage and thus depends on cell type and the genetic predisposition of a cell.

Besides the complexity of damage induced by charged particles, the restriction of damage to defined subnuclear regions in particular at low particle energies represents a unique tool to study the spatial aspects of the cellular response to DNA damage. The importance of these aspects is reflected in the significantly growing interest in microbeam facilities, which allow the directed irradiation of defined subcellular and subnuclear regions with high precision.

The increased relative biological effectiveness of charged particles with respect to, e.g., cell killing is restricted to the lighter particles; for heavier particles, saturation effects lead to a corresponding reduction of RBE. An optimal efficiency of cell killing is observed for particles in the range from helium to carbon at low energies and can be exploited in charged particle tumor therapy. However, due to the complex dependencies of RBE on tissue type and biological endpoint, this optimum is not uniquely defined but depends on the particular combination of tumor tissue and normal tissue. Future research will thus focus on individual optimization with respect to the above-mentioned parameters and will accompany the growing interest for applications of charged particle beams in tumor therapy.

Acknowledgement. I would like to thank Claudia Fournier and Gisela Taucher-Scholz for their comments and suggestions to improve the manuscript. For his continuous support during the last years I would link to thank Gerhard Kraft. I also profited from numerous discussions with my colleagues at the GSI Biophysics group about the physical and in particular the biological aspects of radiation action.

References

1. Hall EJ, Freyer GA (1991) Basic Life Sci 58:3
2. Gademann G (1993) Proceedings of the 1st Symposium on Hadrontherapy, Como, Italy, October 18–21, p 59
3. Wilson RR (1946) Radiology 47:487
4. Griffin TW, Wambersie A, Laramore G, Castro J (1988) Int J Radiat Oncol Biol Phys 14 Suppl 1:S83
5. Blakely EA, Ngo FQH, Curst SB, Tobias CA (1984) In: Lett JT (ed) Advances in radiation biology, vol 11. Academic Press, New York, p 295
6. Kraft G (1990) Strahlenther Onkol 166:10
7. Kraft G (1998) Tumori 84:200
8. Yang TC, Craise LM, Mei MT, Tobias CA (1989) Adv Space Res 9:131
9. Schimmerling W, Wilson JW, Cucinotta F, Kim MH (1998) Phys Med 14 Suppl 1:29
10. Hutchinson C, Glover DM (1995) Cell cycle control. IRL Press, Oxford
11. Bruzzone R, White TW, Goodenough DA (1996) Bioessays 18:709

12. Mothersill C, Seymour C (2000) Radiats Biol Radioecol 40:615
13. Barcellos-Hoff MH (1998) Radiat Res 150:S109
14. Barcellos-Hoff MH, Brooks AL (2001) Radiat Res 156:618
15. Altman KI, Gerber GB, Okada S (1970) Radiation chemistry, vol I: cells. Academic Press, New York
16. Munro TR (1970) Radiat Res 42:451
17. Frankenberg D, Frankenberg-Schwager M, Blöcher D, Harbich R (1981) Radiat Res 88:524
18. Frankenberg-Schwager M, Frankenberg D (1990) Int J Radiat Biol 58:569
19. Iliakis G (1991) Bioessays 13:641
20. Newman HC, Prise KM, Michael BD (2000) Int J Radiat Biol 76:1085
21. Jenner TJ, Cunniffe SM, Stevens DL, O'Neill P (1998) Radiat Res 150:593
22. Jenner TJ, Fulford J, O'Neill P (2001) Radiat Res 156:590
23. Jackson SP (2001) Biochem Soc Trans 29:655
24. Khanna KK, Jackson SP (2001) Nat Genet 27:247–54
25. Prise KM (1994) Int J Radiat Biol 65:43
26. Olive PL (1998) Radiat Res 150:S42
27. Nunez MI, McMillan TJ, Valenzuela MT, Ruiz de Almodovar JM, Pedraza V (1996) Radiother Oncol 39:155
28. Taucher-Scholz G, Heilmann J, Schneider M, Kraft G (1995) Radiat Environ Biophys 34:101
29. Müller WU, Streffer C (1991) Int J Radiat Biol 59:863
30. Natarajan AT (2001) J Environ Pathol Toxicol Oncol 20:293
31. Ritter S, Nasonova E, Scholz M, Kraft-Weyrather W, Kraft G (1996) Int J Radiat Biol 69:155
32. Scholz M, Ritter S, Kraft G (1998) Int J Radiat Biol 74:325
33. Durante M, Furusawa Y, George K, Gialanella G, Greco O, Grossi G, Matsufuji N, Pugliese M, Yang TC (1998) Radiat Res 149:446
34. Nasonova E, Gudowska-Nowak E, Ritter S, Kraft G (2001) Int J Radiat Biol 77:59
35. Puck TT, Markus PI (1956) J Exp Med 103:653
36. Jeggo PA (1990) Mutat Res 239:1
37. Prise KM, Ahnström G, Belli M, Carlsson J, Frankenberg D, Kiefer J, Löbrich M, Michael BD, Nygren J, Simone G, Stenerlöw B (1998) Int J Radiat Biol 74:173–84
38. Venter JC et al. (2001) Science 291:1304
39. Lander ES et al. (2001) Nature 409:860
40. Skarsgard LD, Harrison I, Durand RE (1991) Radiat Res 127:248
41. Shenoy MA, Singh BB (1992) Cancer Invest 10:533
42. Gregoire V, Hittelman WN, Rosier JF, Milas L (1999) Oncol Rep 6:949
43. Maisin JR (1998) Int J Radiat Biol 73:443
44. Wasserman TH, Brizel DM (2001) Oncology (Huntingt) 15:1349
45. Thacker J (1986) Int J Radiat Biol Relat Stud Phys Chem Med 50:1
46. Hagen U (1990) Radiat Environ Biophys 29:315
47. Chen DJ, Tsuboi K, Nguyen T, Yang TC (1994) Adv Space Res 14:347
48. Kiefer J, Schreiber A, Gutermuth F, Koch S, Schmidt P (1999) Mutat Res 431:429
49. Rodemann HP, Bamberg M (1995) Radiother Oncol 35:83
50. Fournier C, Scholz M, Weyrather WK, Rodemann HP, Kraft G (2001) Int J Radiat Biol 77:713
51. Damia G, Imperatori L, Stefanini M, D'Incalci M (1996) Int J Cancer 66:779
52. Singleton BK, Priestley A, Steingrimsdottir H, Gell D, Blunt T, Jackson SP, Lehmann AR, Jeggo PA (1997) Mol Cell Biol 17:1264
53. Withers HR, Taylor JM, Maciejewski B (1988) Int J Radiat Oncol Biol Phys 14:751
54. Constine LS (1991) Pediatrician 18:37
55. Kummermehr J, Trott KR (1996) In: Potten C (ed) Stem cells. Academic Press, London, p 363

56. Durand RE (1990) Cell Tissue Kinet 23:141
57. Santini MT, Rainaldi G, Indovina PL (1999) Int J Radiat Biol 75:787
58. Müller-Klieser W (2000) Crit Rev Oncol Hematol 36:123
59. Nakazawa K, Kalassy M, Sahuc F, Collombel C, Damour O (1998) Med Biol Eng Comput 36:813
60. Krämer M, Kraft G (1994) Radiat Environ Biophys 33:91
61. Ramm U, Bechthold U, Jagutzki O, Hagmann S, Kraft G, Schmidt-Böcking H (1995) Nucl Instr Meth B 98:359–362
62. Kiefer J, Straaten H (1986) Phys Med Biol 31:1201
63. Bethe HA (1930) Ann Phys 5:325
64. Bloch F (1933) Ann Phys 16:287
65. Heinrich W, Wiegel B, Kraft G (1991) GSI-Preprint GSI-91–30
66. Caswell RS, Coyne JJ (1972) Radiat Res 52:448
67. Morstin K, Dydejczk A, Booz J (1985) In: Nuclear and atomic data for radiotherapy and related radiobiology. Proceedings of Advisory Group Meeting, Rijswijk, IAEA Vienna, p 239
68. Zirkle RE (1935) Am J Cancer 28:558
69. Giles NH, Tobias CA (1954) Science 120:993
70. Barendsen GW, Beusker TLJ, Vergroesen AJ, Budke L (1960) Radiat Res 13:841
71. Barendsen GW, Walter HMD, Fowler JF, Bewley DK (1963) Radiat Res 18:106
72. Weyrather WK, Ritter S, Scholz M, Kraft G (1999) Int J Radiat Biol 75:1357
73. Goodhead DT, Belli M, Mill AJ, Bance DA, Allen LA, Hall SC, Ianzini F, Simone G, Stevens DL, Stretch A et al. (1992) Int J Radiat Biol 61:611
74. Belli M, Cera F, Cherubini R, Haque AM, Ianzini F, Moschini G, Sapora O, Simone G, Tabocchini MA, Tiveron P (1993) Int J Radiat Biol 63:331
75. Furusawa Y, Fukutsu K, Aoki M, Itsukaichi H, Eguchi-Kasai K, Ohara H, Yatagai F, Kanai T, Ando K (2000) Radiat Res 154:485
76. Jeggo PA, Kemp LM (1983) Mutat Res 112:313
77. Jeggo PA (1985) Mutat Res 146:265
78. Kraft G (1987) Nucl Sci Appl 3:1
79. Blakely EA, Tobias CA, Yang TC, Smith KC, Lyman JT (1979) Radiat Res 80:122
80. Barendsen GW, Koot CJ, van Kersen GR, Bewley DK, Field SB, Parnell CJ (1966) Int J Radiat Biol 10:317
81. Taucher-Scholz G, Heilmann J, Kraft G (1996) Adv Space Res 18:83
82. Rothkamm K, Kühne M, Jeggo PA, Löbrich M (2001) Cancer Res 61:1886
83. Ritter S, Kraft-Weyrather W, Scholz M, Kraft G (1992) Adv Space Res 12:119
84. Zorn C, Cremer T, Cremer C, Zimmer J (1976) Hum Genet 35:83
85. Tobias CA, Blakely EA, Alpen EL, Castro JR, Ainsworth EJ, Curtis SB, Ngo FQ, Rodriguez A, Roots RJ, Tenforde T, Yang TC (1982) Int J Radiat Oncol Biol Phys 8:2109
86. Jakob B, Scholz M, Taucher-Scholz G (2000) Radiat Res 154:398
87. Jakob B, Scholz M, Taucher-Scholz G (2002) Int J Radiat Biol 78:75
88. Scholz M, Jakob B, Taucher-Scholz G (2001) Radiat Res 156:558
89. Folkard M, Vojnovic B, Prise KM, Bowey AG, Locke RJ, Schettino G, Michael BD (1997) Int J Radiat Biol 72:375
90. Folkard M, Vojnovic B, Hollis KJ, Bowey AG, Watts SJ, Schettino G, Prise KM, Michael BD (1997) Int J Radiat Biol 72:387
91. Randers-Pehrson G, Geard CR, Johnson G, Elliston CD, Brenner DJ (2001) Radiat Res 156:210
92. Prise KM, Belyakov OV, Folkard M, Michael BD (1998) Int J Radiat Biol 74:793–8
93. Sawant SG, Randers-Pehrson G, Metting NF, Hall EJ (2001) Radiat Res 156:177
94. Zhou H, Randers-Pehrson G, Waldren CA, Vannais D, Hall EJ, Hei TK (2000) Proc Natl Acad Sci U S A 97:2099
95. Sawant SG, Randers-Pehrson G, Geard CR, Brenner DJ, Hall EJ (2001) Radiat Res 155:397

96. Azzam EI, de Toledo SM, Gooding T, Little JB (1998) Radiat Res 150:497
97. Azzam EI, de Toledo SM, Little JB (2001) Proc Natl Acad Sci USA 98:473
98. Tsoulou E, Baggio L, Cherubini R, Kalfas CA (2001) Int J Radiat Biol 77:1133
99. Schettino G, Folkard M, Prise KM, Vojnovic B, Bowey AG, Michael BD (2001) Radiat Res 156:526
100. Böhrnsen G, Weber KJ, Scholz M (2002) Int J Radiat Biol 78:259
101. Belyakov OV, Folkard M, Mothersill C, Prise KM, Michael BD (2001) Abstract, 48th Meeting of the Radiation Research Society, San Juan, Puerto Rico
102. Alpen EL, Powers-Risius P, McDonald M (1980) Radiat Res 83:677
103. Leith JT (1983) In: Lett J (ed) Advances in radiation biology, vol 10, p 191
104. Rodriguez A, Alpen EL, Powers-Risius P (1992) Radiat Res 132:184
105. Zacharias T, Dörr W, Enghardt W, Haberer T, Krämer M, Kumpf R, Rothig H, Scholz M, Weber U, Kraft G, Herrmann T (1997) Acta Oncol 36:637
106. Fukutsu K, Kanai T, Furusawa Y, Ando K (1997) Radiat Res 148:168
107. Ando K, Koike S, Nojima K, Chen YJ, Ohira C, Ando S, Kobayashi N, Ohbuchi T, Shimizu W, Kanai T (1998) Int J Radiat Biol 74:129
108. Dörr W, Alheit H, Appold S, Enghardt W, Haase M, Haberer T, Hinz R, Jäkel O, Kellerer AM, Krämer M, Kraft G, Kumpf R, Nitzsche H, Scholz M, Voigtmann L, Herrmann T (1999) Radiat Environ Biophys 38:185
109. Okada S, Okeda R, Matsushita S, Kawano A (1998) Radiat Res 150:304
110. Tomizawa M, Miyamoto T, Kato H, Otsu H (2000) J Radiat Res (Tokyo) 41:151
111. Inouye M, Takahashi S, Kubota Y, Hayasaka S, Murata Y (2000) J Radiat Res (Tokyo) 41:303
112. Hopewell JW, Barnes DW, Goodhead DT, Knowles JF, Wiernik G, Young CM (1982) Int J Radiat Oncol Biol Phys 8:2077
113. Hornsey S (1982) Int J Radiat Oncol Biol Phys 8:2099
114. Withers HR, Thames HD Jr, Peters LJ (1982) Int J Radiat Oncol Biol Phys 8:2071
115. Curtis SB, Tenforde TS, Parks D, Schilling WA, Lyman JT (1978) Radiat Res 74:274
116. Curtis SB, Tenforde TS (1980) Br J Cancer Suppl 41:266
117. Tenforde TS, Curtis SB, Crabtree KE, Tenforde SD, Schilling WA, Howard J, Lyman JT (1980) Radiat Res 83:42
118. Tenforde TS, Tenforde SD, Crabtree KE, Parks DL, Schilling WA, Parr SS, Flynn MJ, Howard J, Lyman JT, Curtis SB (1981) Int J Radiat Oncol Biol Phys 7:217
119. Afzal SM, Tenforde TS, Kavanau KS, Curtis SB (1991) Radiat Res 127:230
120. Ando K, Koike S, Ohira C, Chen YJ, Nojima K, Ando S, Ohbuchi T, Kobayashi N, Shimizu W, Urano M (1999) Int J Radiat Biol 75:505
121. Oya N, Sasai K, Shibata T, Takagi T, Shibuya K, Koike S, Nojima K, Furusawa Y, Ando K, Hiraoka M (2001) J Radiat Res (Tokyo) 42:131
122. Neary GJ (1965) Mutat Res 2:242
123. Kellerer AM, Rossi HH (1972) Curr Top Radiat Res Q 8:85
124. Rossi HH, Zaider M (1996) Microdosimetry. Springer, Berlin Heidelberg New York
125. Goodhead DT (1977) Int J Radiat Biol 32:43
126. Kellerer AM, Rossi HH (1978) Radiat Res 75:471
127. Rossi HH, Zaider M (1992) Radiat Res 132:178
128. Ward JF (1994) Int J Radiat Biol 66:427
129. Heilmann J, Rink H, Taucher-Scholz G, Kraft G (1993) Radiat Res 135:46
130. Heilmann J, Taucher-Scholz G, Kraft G (1995) Int J Radiat Biol 68:153
131. Weber KJ, Flentje M (1993) Int J Radiat Biol 64:169
132. Michalik V (1991) Phys Med Biol 36:1001
133. Michalik V (1993) Radiat Res 134:265
134. Ottolenghi A, Merzagora M, Paretzke HG (1997) Radiat Environ Biophys 36:97
135. Nikjoo H, Uehara S, Wilson WE, Hoshi M, Goodhead DT (1998) Int J Radiat Biol 73:355
136. Ottolenghi A, Monforti F, Merzagora M (1997) Int J Radiat Biol 72:505

137. Butts JJ, Katz R (1967) Radiat Res 30:855
138. Katz R, Ackerson B, Homayoonfar M, Sharma SC (1971) Radiat Res 47:402
139. Katz R, Dunn DE, Sinclair GL (1985) Rad Prot Dosim 13:281
140. Scholz M, Kraft G (1996) Adv Space Res 18:5
141. Scholz M (1996) Bull Cancer Radiother 83 Suppl:50 s
142. Scholz M, Kellerer AM, Kraft-Weyrather W, Kraft G (1997) Radiat Environ Biophys 36:59
143. Kellerer AM (1998) Physica Medica 14 Suppl 1:48
144. Sachs RK, Chen AM, Brenner DJ (1997) Int J Radiat Biol 71:1
145. Hahnfeldt P, Hearst JE, Brenner DJ, Sachs RK, Hlatky LR (1993) Proc Natl Acad Sci USA 90:7854
146. Cremer C, Munkel C, Granzow M, Jauch A, Dietzel S, Eils R, Guan XY, Meltzer PS, Trent JM, Langowski J, Cremer T (1996) Mutat Res 366:97
147. Munkel C, Eils R, Dietzel S, Zink D, Mehring C, Wedemann G, Cremer T, Langowski J (1999) J Mol Biol 285:1053
148. Goitein M, Chen GT (1983) Med Phys 10:831
149. Blattmann H (1992) Radiat Environ Biophys 31:219
150. Haberer T, Becher W, Schardt D, Kraft G (1993) Nucl Instr Meth A 330:296
151. Eickhoff H, Haberer T, Kraft G, Krause U, Richter M, Steiner R, Debus J (1999) Strahlenther Onkol 175 Suppl 2:21
152. Blattmann H, Coray A, Pedroni E, Greiner R (1990) Strahlenther Onkol 166:45
153. Pedroni E, Bacher R, Blattmann H, Bohringer T, Coray A, Lomax A, Lin S, Munkel G, Scheib S, Schneider U et al. (1995) Med Phys 22:37
154. Enghardt W, Debus J, Haberer T, Hasch BG, Hinz R, Jäkel O, Krämer M, Lauckner K, Pawelke J (1999) Strahlenther Onkol 175 Suppl 2:33
155. Austin-Seymour M, Munzenrider JE, Goitein M, Gentry R, Gragoudas E, Koehler AM, McNulty P, Osborne E, Ryugo DK, Seddon J et al. (1985) Radiat Res Suppl 8:S219
156. Wouters BG, Lam GK, Oelfke U, Gardey K, Durand RE, Skarsgard LD (1996) Radiat Res 146:159
157. Pihet P, Menzel HG, Schmidt R, Beauduin M, Wambersie A (1990) Radiat Prot Dosim 31:437
158. Loncol T, Cosgrove V, Denis JM, Gueulette J, Mazal A, Menzel HG, Pihet P, Sabatier R (1994) Radiat Prot Dosim 52:347
159. Kanai T, Endo M, Minohara S, Miyahara N, Koyama-ito H, Tomura H, Matsufuji N, Futami Y, Fukumura A, Hiraoka T, Furusawa Y, Ando K, Suzuki M, Soga F, Kawachi K (1999) Int J Radiat Oncol Biol Phys 44:201
160. Krämer M, Scholz M (2000) Phys Med Biol 45:3319
161. Suit H, Urie M (1992) J Natl Cancer Inst 84:155
162. Castro JR (1995) Radiat Environ Biophys 34:45
163. Kraft G (2000) Prog Part Nucl Phys 45:S473
164. Debus J, Haberer T, Schulz-Ertner D, Jäkel O, Wenz F, Enghardt W, Schlegel W, Kraft G, Wannenmacher M (2000) Strahlenther Onkol 176:211

Author Index Volumes 101–162

Author Index Volumes 1–100 see Volume 100

de, Abajo, J. and de la Campa, J.G.: Processable Aromatic Polyimides. Vol. 140, pp. 23-60.

Adolf, D. B. see Ediger, M. D.: Vol. 116, pp. 73-110.

Aharoni, S. M. and Edwards, S. F.: Rigid Polymer Networks. Vol. 118, pp. 1-231.

Albertsson, A.-C., Varma, I. K.: Aliphatic Polyesters: Synthesis, Properties and Applications. Vol. 157, pp. 99–138.

Albertsson, A.-C. see Edlund, U.: Vol. 157, pp. 53-98.

Albertsson, A.-C. see Söderqvist Lindblad, M.: Vol. 157, pp. 139–161.

Albertsson, A.-C. see Stridsberg, K. M.: Vol. 157, pp. 27–51.

Améduri, B., Boutevin, B. and Gramain, P.: Synthesis of Block Copolymers by Radical Polymerization and Telomerization. Vol. 127, pp. 87-142.

Améduri, B. and Boutevin, B.: Synthesis and Properties of Fluorinated Telechelic Monodispersed Compounds. Vol. 102, pp. 133-170.

Amselem, S. see Domb, A. J.: Vol. 107, pp. 93-142.

Andrady, A. L.: Wavelenght Sensitivity in Polymer Photodegradation. Vol. 128, pp. 47-94.

Andreis, M. and Koenig, J. L.: Application of Nitrogen-15 NMR to Polymers. Vol. 124, pp. 191-238.

Angiolini, L. see Carlini, C.: Vol. 123, pp. 127-214.

Anjum, N. see Gupta, B.: Vol. 162, pp. 37-63.

Anseth, K. S., Newman, S. M. and Bowman, C. N.: Polymeric Dental Composites: Properties and Reaction Behavior of Multimethacrylate Dental Restorations. Vol. 122, pp. 177-218.

Antonietti, M. see Cölfen, H.: Vol. 150, pp. 67-187.

Armitage, B. A. see O'Brien, D. F.: Vol. 126, pp. 53-58.

Arndt, M. see Kaminski, W.: Vol. 127, pp. 143-187.

Arnold Jr., F. E. and Arnold, F. E.: Rigid-Rod Polymers and Molecular Composites. Vol. 117, pp. 257-296.

Arora, M. see Kumar, M.N.V.R.: Vol. 160, pp. 45-118.

Arshady, R.: Polymer Synthesis via Activated Esters: A New Dimension of Creativity in Macromolecular Chemistry. Vol. 111, pp. 1-42.

Bahar, I., Erman, B. and Monnerie, L.: Effect of Molecular Structure on Local Chain Dynamics: Analytical Approaches and Computational Methods. Vol. 116, pp. 145-206.

Ballauff, M. see Dingenouts, N.: Vol. 144, pp. 1-48.

Baltá-Calleja, F. J., González Arche, A., Ezquerra, T. A., Santa Cruz, C., Batallón, F., Frick, B. and López Cabarcos, E.: Structure and Properties of Ferroelectric Copolymers of Poly(vinylidene) Fluoride. Vol. 108, pp. 1-48.

Barnes, M. D. see Otaigbe, J.U.: Vol. 154, pp. 1-86.

Barshtein, G. R. and Sabsai, O. Y.: Compositions with Mineralorganic Fillers. Vol. 101, pp.1-28.

Baschnagel, J., Binder, K., Doruker, P., Gusev, A. A., Hahn, O., Kremer, K., Mattice, W. L., Müller-Plathe, F., Murat, M., Paul, W., Santos, S., Sutter, U. W., Tries, V.: Bridging the Gap Between Atomistic and Coarse-Grained Models of Polymers: Status and Perspectives. Vol. 152, pp. 41-156.

Batallán, F. see Baltá-Calleja, F. J.: Vol. 108, pp. 1-48.

Batog, A. E., Pet'ko, I. P., Penczek, P.: Aliphatic-Cycloaliphatic Epoxy Compounds and Polymers. Vol. 144, pp. 49-114.

Barton, J. see Hunkeler, D.: Vol. 112, pp. 115-134.

Bell, C. L. and *Peppas, N. A.*: Biomedical Membranes from Hydrogels and Interpolymer Complexes. Vol. 122, pp. 125-176.

Bellon-Maurel, A. see Calmon-Decriaud, A.: Vol. 135, pp. 207-226.

Bennett, D. E. see O'Brien, D. F.: Vol. 126, pp. 53-84.

Berry, G.C.: Static and Dynamic Light Scattering on Moderately Concentraded Solutions: Isotropic Solutions of Flexible and Rodlike Chains and Nematic Solutions of Rodlike Chains. Vol. 114, pp. 233-290.

Bershtein, V. A. and *Ryzhov, V. A.*: Far Infrared Spectroscopy of Polymers. Vol. 114, pp. 43-122.

Bigg, D. M.: Thermal Conductivity of Heterophase Polymer Compositions. Vol. 119, pp. 1-30.

Binder, K.: Phase Transitions in Polymer Blends and Block Copolymer Melts: Some Recent Developments. Vol. 112, pp. 115-134.

Binder, K.: Phase Transitions of Polymer Blends and Block Copolymer Melts in Thin Films. Vol. 138, pp. 1-90.

Binder, K. see Baschnagel, J.: Vol. 152, pp. 41-156.

Bird, R. B. see Curtiss, C. F.: Vol. 125, pp. 1-102.

Biswas, M. and *Mukherjee, A.*: Synthesis and Evaluation of Metal-Containing Polymers. Vol. 115, pp. 89-124.

Biswas, M. and *Sinha Ray, S.*: Recent Progress in Synthesis and Evaluation of Polymer-Montmorillonite Nanocomposites. Vol. 155, pp. 167-221.

Bolze, J. see Dingenouts, N.: Vol. 144, pp. 1-48.

Bosshard, C.: see Gubler, U.: Vol. 158, pp. 123-190.

Boutevin, B. and *Robin, J. J.*: Synthesis and Properties of Fluorinated Diols. Vol. 102. pp. 105-132.

Boutevin, B. see Amédouri, B.: Vol. 102, pp. 133-170.

Boutevin, B. see Améduri, B.: Vol. 127, pp. 87-142.

Bowman, C. N. see Anseth, K. S.: Vol. 122, pp. 177-218.

Boyd, R. H.: Prediction of Polymer Crystal Structures and Properties. Vol. 116, pp. 1-26.

Briber, R. M. see Hedrick, J. L.: Vol. 141, pp. 1-44.

Bronnikov, S. V., Vettegren, V. I. and *Frenkel, S. Y.*: Kinetics of Deformation and Relaxation in Highly Oriented Polymers. Vol. 125, pp. 103-146.

Brown, H. R. see Creton, C.: Vol. 156, pp. 53-135.

Bruza, K. J. see Kirchhoff, R. A.: Vol. 117, pp. 1-66.

Budkowski, A.: Interfacial Phenomena in Thin Polymer Films: Phase Coexistence and Segregation. Vol. 148, pp. 1-112.

Burban, J. H. see Cussler, E. L.: Vol. 110, pp. 67-80.

Burchard, W.: Solution Properties of Branched Macromolecules. Vol. 143, pp. 113-194.

Calmon-Decriaud, A. Bellon-Maurel, V., Silvestre, F.: Standard Methods for Testing the Aerobic Biodegradation of Polymeric Materials. Vol 135, pp. 207-226.

Cameron, N. R. and *Sherrington, D. C.*: High Internal Phase Emulsions (HIPEs)-Structure, Properties and Use in Polymer Preparation. Vol. 126, pp. 163-214.

de la Campa, J. G. see de Abajo, , J.: Vol. 140, pp. 23-60.

Candau, F. see Hunkeler, D.: Vol. 112, pp. 115-134.

Canelas, D. A. and *DeSimone, J. M.*: Polymerizations in Liquid and Supercritical Carbon Dioxide. Vol. 133, pp. 103-140.

Canva, M., Stegeman, G. I.: Quadratic Parametric Interactions in Organic Waveguides. Vol. 158, pp. 87-121.

Capek, I.: Kinetics of the Free-Radical Emulsion Polymerization of Vinyl Chloride. Vol. 120, pp. 135-206.

Capek, I.: Radical Polymerization of Polyoxyethylene Macromonomers in Disperse Systems. Vol. 145, pp. 1-56.

Capek, I.: Radical Polymerization of Polyoxyethylene Macromonomers in Disperse Systems. Vol. 146, pp. 1-56.

Capek, I. and *Chern, C.-S.:* Radical Polymerization in Direct Mini-Emulsion Systems. Vol. 155, pp. 101-166.

Carlesso, G. see Prokop, A.: Vol. 160, pp. 119–174.

Carlini, C. and *Angiolini, L.:* Polymers as Free Radical Photoinitiators. Vol. 123, pp. 127-214.

Carter, K. R. see Hedrick, J. L.: Vol. 141, pp. 1-44.

Casas-Vazquez, J. see Jou, D.: Vol. 120, pp. 207-266.

Chandrasekhar, V.: Polymer Solid Electrolytes: Synthesis and Structure. Vol 135, pp. 139-206.

Chang, J. Y. see Han, M. J.: Vol. 153, pp. 1-36.

Charleux, B., Faust R.: Synthesis of Branched Polymers by Cationic Polymerization. Vol. 142, pp. 1-70.

Chen, P. see Jaffe, M.: Vol. 117, pp. 297-328.

Chern, C.-S. see Capek, I.: Vol. 155, pp. 101-166.

Chevolot, Y. see Mathieu, H. J.: Vol. 162, pp. 1-35.

Choe, E.-W. see Jaffe, M.: Vol. 117, pp. 297-328.

Chow, T. S.: Glassy State Relaxation and Deformation in Polymers. Vol. 103, pp. 149-190.

Chung, S.-J. see Lin, T.-C.: Vol. 161, pp. 157-193

Chung, T.-S. see Jaffe, M.: Vol. 117, pp. 297-328.

Cölfen, H. and *Antonietti, M.:* Field-Flow Fractionation Techniques for Polymer and Colloid Analysis. Vol. 150, pp. 67-187.

Comanita, B. see Roovers, J.: Vol. 142, pp. 179-228.

Connell, J. W. see Hergenrother, P. M.: Vol. 117, pp. 67-110.

Creton, C., Kramer, E. J., Brown, H. R., Hui, C.-Y.: Adhesion and Fracture of Interfaces Between Immiscible Polymers: From the Molecular to the Continuum Scale. Vol. 156, pp. 53-135.

Criado-Sancho, M. see Jou, D.: Vol. 120, pp. 207-266.

Curro, J.G. see Schweizer, K.S.: Vol. 116, pp. 319-378.

Curtiss, C. F. and *Bird, R. B.:* Statistical Mechanics of Transport Phenomena: Polymeric Liquid Mixtures. Vol. 125, pp. 1-102.

Cussler, E. L., Wang, K. L. and *Burban, J. H.:* Hydrogels as Separation Agents. Vol. 110, pp. 67-80.

Dalton, L. Nonlinear Optical Polymeric Materials: From Chromophore Design to Commercial Applications. Vol. 158, pp. 1-86.

Davidson, J.M. see Prokop, A.: Vol. 160, pp.119–174.

DeSimone, J. M. see Canelas D. A.: Vol. 133, pp. 103-140.

DiMari, S. see Prokop, A.: Vol. 136, pp. 1-52.

Dimonie, M. V. see Hunkeler, D.: Vol. 112, pp. 115-134.

Dingenouts, N., Bolze, J., Pötschke, D., Ballauf, M.: Analysis of Polymer Latexes by Small-Angle X-Ray Scattering. Vol. 144, pp. 1-48.

Dodd, L. R. and *Theodorou, D. N.:* Atomistic Monte Carlo Simulation and Continuum Mean Field Theory of the Structure and Equation of State Properties of Alkane and Polymer Melts. Vol. 116, pp. 249-282.

Doelker, E.: Cellulose Derivatives. Vol. 107, pp. 199-266.

Dolden, J. G.: Calculation of a Mesogenic Index with Emphasis Upon LC-Polyimides. Vol. 141, pp. 189-245.

Domb, A. J., Amselem, S., Shah, J. and *Maniar, M.:* Polyanhydrides: Synthesis and Characterization. Vol.107, pp. 93-142.

Domb, A.J. see Kumar, M.N.V.R.: Vol. 160, pp. 45–118.

Doruker, P. see Baschnagel, J.: Vol. 152, pp. 41-156.

Dubois, P. see Mecerreyes, D.: Vol. 147, pp. 1-60.

Dubrovskii, S. A. see Kazanskii, K. S.: Vol. 104, pp. 97-134.

Dunkin, I. R. see Steinke, J.: Vol. 123, pp. 81-126.

Dunson, D. L. see McGrath, J. E.: Vol. 140, pp. 61-106.

Eastmond, G. C.: Poly(ε-caprolactone) Blends. Vol.149, pp. 59-223.

Economy, J. and *Goranov, K.:* Thermotropic Liquid Crystalline Polymers for High Performance Applications. Vol. 117, pp. 221-256.

Ediger, M. D. and *Adolf, D. B.*: Brownian Dynamics Simulations of Local Polymer Dynamics. Vol. 116, pp. 73-110.

Edlund, U. Albertsson, A.-C.: Degradable Polymer Microspheres for Controlled Drug Delivery. Vol. 157, pp. 53-98.

Edwards, S. F. see Aharoni, S. M.: Vol. 118, pp. 1-231.

Endo, T. see Yagci, Y.: Vol. 127, pp. 59-86.

Engelhardt, H. and *Grosche, O.*: Capillary Electrophoresis in Polymer Analysis. Vol. 150, pp. 189-217.

Erman, B. see Bahar, I.: Vol. 116, pp. 145-206.

Ewen, B, Richter, D.: Neutron Spin Echo Investigations on the Segmental Dynamics of Polymers in Melts, Networks and Solutions. Vol. 134, pp. 1-130.

Ezquerra, T. A. see Baltá-Calleja, F. J.: Vol. 108, pp. 1-48.

Faust, R. see Charleux, B: Vol. 142, pp. 1-70.

Fekete, E see Pukánszky, B: Vol. 139, pp. 109-154.

Fendler, J.H.: Membrane-Mimetic Approach to Advanced Materials. Vol. 113, pp. 1-209.

Fetters, L. J. see Xu, Z.: Vol. 120, pp. 1-50.

Förster, S. and *Schmidt, M.*: Polyelectrolytes in Solution. Vol. 120, pp. 51-134.

Freire, J. J.: Conformational Properties of Branched Polymers: Theory and Simulations. Vol. 143, pp. 35-112.

Frenkel, S. Y. see Bronnikov, S. V.: Vol. 125, pp. 103-146.

Frick, B. see Baltá-Calleja, F. J.: Vol. 108, pp. 1-48.

Fridman, M. L.: see Terent´eva, J. P.: Vol. 101, pp. 29-64.

Fukui, K. see Otaigbe, J. U.: Vol. 154, pp. 1-86.

Funke, W.: Microgels-Intramolecularly Crosslinked Macromolecules with a Globular Structure. Vol. 136, pp. 137-232.

Galina, H.: Mean-Field Kinetic Modeling of Polymerization: The Smoluchowski Coagulation Equation. Vol. 137, pp. 135-172.

Ganesh, K. see Kishore, K.: Vol. 121, pp. 81-122.

Gaw, K. O. and Kakimoto, M.: Polyimide-Epoxy Composites. Vol. 140, pp. 107-136.

Geckeler, K. E. see Rivas, B.: Vol. 102, pp. 171-188.

Geckeler, K. E.: Soluble Polymer Supports for Liquid-Phase Synthesis. Vol. 121, pp. 31-80.

Gehrke, S. H.: Synthesis, Equilibrium Swelling, Kinetics Permeability and Applications of Environmentally Responsive Gels. Vol. 110, pp. 81-144.

de Gennes, P.-G.: Flexible Polymers in Nanopores. Vol. 138, pp. 91-106.

Giannelis, E.P., Krishnamoorti, R., Manias, E.: Polymer-Silicate Nanocomposites: Model Systems for Confined Polymers and Polymer Brushes. Vol. 138, pp. 107-148.

Godovsky, D. Y.: Device Applications of Polymer-Nanocomposites. Vol. 153, pp. 163-205.

Godovsky, D. Y.: Electron Behavior and Magnetic Properties Polymer-Nanocomposites. Vol. 119, pp. 79-122.

González Arche, A. see Baltá-Calleja, F. J.: Vol. 108, pp. 1-48.

Goranov, K. see Economy, J.: Vol. 117, pp. 221-256.

Gramain, P. see Améduri, B.: Vol. 127, pp. 87-142.

Grest, G.S.: Normal and Shear Forces Between Polymer Brushes. Vol. 138, pp. 149-184.

Grigorescu, G, Kulicke, W.-M.: Prediction of Viscoelastic Properties and Shear Stability of Polymers in Solution. Vol. 152, p. 1-40.

Grosberg, A. and Nechaev, S.: Polymer Topology. Vol. 106, pp. 1-30.

Grosche, O. see Engelhardt, H.: Vol. 150, pp. 189-217.

Grubbs, R., Risse, W. and *Novac, B.*: The Development of Well-defined Catalysts for Ring-Opening Olefin Metathesis. Vol. 102, pp. 47-72.

Gubler, U., Bosshard, C.: Molecular Design for Third-Order Nonlinear Optics. Vol. 158, pp. 123-190.

van Gunsteren, W. F. see Gusev, A. A.: Vol. 116, pp. 207-248.

Gupta, B., Anjum, N.: Plasma and Radiation-Induced Graft Modification of Polymers for Biomedical Applications. Vol. 162, pp. 37-63.

Gusev, A. A., Müller-Plathe, F., van Gunsteren, W. F. and *Suter, U. W.*: Dynamics of Small Molecules in Bulk Polymers. Vol. 116, pp. 207-248.
Gusev, A. A. see Baschnagel, J.: Vol. 152, pp. 41-156.
Guillot, J. see Hunkeler, D.: Vol. 112, pp. 115-134.
Guyot, A. and *Tauer, K.*: Reactive Surfactants in Emulsion Polymerization. Vol. 111, pp. 43-66.

Hadjichristidis, N., Pispas, S., Pitsikalis, M., Iatrou, H., Vlahos, C.: Asymmetric Star Polymers Synthesis and Properties. Vol. 142, pp. 71-128.
Hadjichristidis, N. see Xu, Z.: Vol. 120, pp. 1-50.
Hadjichristidis, N. see Pitsikalis, M.: Vol. 135, pp. 1-138.
Hahn, O. see Baschnagel, J.: Vol. 152, pp. 41-156.
Hakkarainen, M.: Aliphatic Polyesters: Abiotic and Biotic Degradation and Degradation Products. Vol. 157, pp. 1-26.
Hall, H. K. see Penelle, J.: Vol. 102, pp. 73-104.
Hamley, I. W.: Crystallization in Block Copolymers. Vol. 148, pp. 113-138.
Hammouda, B.: SANS from Homogeneous Polymer Mixtures: A Unified Overview. Vol. 106, pp. 87-134.
Han, M.J. and *Chang, J.Y.*: Polynucleotide Analogues. Vol. 153, pp. 1-36.
Harada, A.: Design and Construction of Supramolecular Architectures Consisting of Cyclodextrins and Polymers. Vol. 133, pp. 141-192.
Haralson, M. A. see Prokop, A.: Vol. 136, pp. 1-52.
Hassan, C.M. and *Peppas, N.A.*: Structure and Applications of Poly(vinyl alcohol) Hydrogels Produced by Conventional Crosslinking or by Freezing/Thawing Methods. Vol. 153, pp. 37-65.
Hawker, C. J. Dentritic and Hyperbranched Macromolecules – Precisely Controlled Macromolecular Architectures. Vol. 147, pp. 113-160.
Hawker, C. J. see Hedrick, J. L.: Vol. 141, pp. 1-44.
He, G. S. see Lin, T.-C.: Vol. 161, pp. 157-193.
Hedrick, J. L., Carter, K. R., Labadie, J. W., Miller, R. D., Volksen, W., Hawker, C. J., Yoon, D. Y., Russell, T. P., McGrath, J. E., Briber, R. M.: Nanoporous Polyimides. Vol. 141, pp. 1-44.
Hedrick, J. L., Labadie, J. W., Volksen, W. and *Hilborn, J. G.*: Nanoscopically Engineered Polyimides. Vol. 147, pp. 61-112.
Hedrick, J. L. see Hergenrother, P. M.: Vol. 117, pp. 67-110.
Hedrick, J. L. see Kiefer, J.: Vol. 147, pp. 161-247.
Hedrick, J.L. see McGrath, J. E.: Vol. 140, pp. 61-106.
Heinrich, G. and *Klüppel, M.*: Recent Advances in the Theory of Filler Networking in Elastomers. Vol. 160, pp. 1-44.
Heller, J.: Poly (Ortho Esters). Vol. 107, pp. 41-92.
Hemielec, A. A. see Hunkeler, D.: Vol. 112, pp. 115-134.
Hergenrother, P. M., Connell, J. W., Labadie, J. W. and *Hedrick, J. L.*: Poly(arylene ether)s Containing Heterocyclic Units. Vol. 117, pp. 67-110.
Hernández-Barajas, J. see Wandrey, C.: Vol. 145, pp. 123-182.
Hervet, H. see Léger, L.: Vol. 138, pp. 185-226.
Hilborn, J. G. see Hedrick, J. L.: Vol. 147, pp. 61-112.
Hilborn, J. G. see Kiefer, J.: Vol. 147, pp. 161-247.
Hiramatsu, N. see Matsushige, M.: Vol. 125, pp. 147-186.
Hirasa, O. see Suzuki, M.: Vol. 110, pp. 241-262.
Hirotsu, S.: Coexistence of Phases and the Nature of First-Order Transition in Poly-N-isopropylacrylamide Gels. Vol. 110, pp. 1-26.
Höcker, H. see Klee, D.: Vol. 149, pp. 1-57.
Hornsby, P.: Rheology, Compoundind and Processing of Filled Thermoplastics. Vol. 139, pp. 155-216.
Hui, C.-Y. see Creton, C.: Vol. 156, pp. 53-135
Hult, A., Johansson, M., Malmström, E.: Hyperbranched Polymers. Vol. 143, pp. 1-34.

Hunkeler, D., Candau, F., Pichot, C., Hemielec, A. E., Xie, T. Y., Barton, J., Vaskova, V., Guillot, J., Dimonie, M. V., Reichert, K. H.: Heterophase Polymerization: A Physical and Kinetic Comparision and Categorization. Vol. 112, pp. 115-134.
Hunkeler, D. see Prokop, A.: Vol. 136, pp. 1-52; 53-74.
Hunkeler, D see Wandrey, C.: Vol. 145, pp. 123-182.

Iatrou, H. see Hadjichristidis, N.: Vol. 142, pp. 71-128.
Ichikawa, T. see Yoshida, H.: Vol. 105, pp. 3-36.
Ihara, E. see Yasuda, H.: Vol. 133, pp. 53-102.
Ikada, Y. see Uyama, Y.: Vol. 137, pp. 1-40.
Ilavsky, M.: Effect on Phase Transition on Swelling and Mechanical Behavior of Synthetic Hydrogels. Vol. 109, pp. 173-206.
Imai, Y.: Rapid Synthesis of Polyimides from Nylon-Salt Monomers. Vol. 140, pp. 1-23.
Inomata, H. see Saito, S.: Vol. 106, pp. 207-232.
Inoue, S. see Sugimoto, H.: Vol. 146, pp. 39-120.
Irie, M.: Stimuli-Responsive Poly(N-isopropylacrylamide), Photo- and Chemical-Induced Phase Transitions. Vol. 110, pp. 49-66.
Ise, N. see Matsuoka, H.: Vol. 114, pp. 187-232.
Ito, K., Kawaguchi, S,:Poly(macronomers), Homo- and Copolymerization. Vol. 142, pp. 129-178.
Ivanov, A. E. see Zubov, V. P.: Vol. 104, pp. 135-176.

Jacob, S. and Kennedy, J.: Synthesis, Characterization and Properties of OCTA-ARM Polyisobutylene-Based Star Polymers. Vol. 146, pp. 1-38.
Jaffe, M., Chen, P., Choe, E.-W., Chung, T.-S. and Makhija, S.: High Performance Polymer Blends. Vol. 117, pp. 297-328.
Jancar, J.: Structure-Property Relationships in Thermoplastic Matrices. Vol. 139, pp. 1-66.
Jen, A. K-Y. see Kajzar, F.: Vol. 161, pp. 1-85.
Jerôme, R.: see Mecerreyes, D.: Vol. 147, pp. 1-60.
Jiang, M., Li, M., Xiang, M. and Zhou, H.: Interpolymer Complexation and Miscibility and Enhancement by Hydrogen Bonding. Vol. 146, pp. 121-194.
Jin, J.: see Shim, H.-K.: Vol. 158, pp. 191-241.
Jo, W. H. and Yang, J. S.: Molecular Simulation Approaches for Multiphase Polymer Systems. Vol. 156, pp. 1-52.
Johansson, M. see Hult, A.: Vol. 143, pp. 1-34.
Joos-Müller, B. see Funke, W.: Vol. 136, pp. 137-232.
Jou, D., Casas-Vazquez, J. and Criado-Sancho, M.: Thermodynamics of Polymer Solutions under Flow: Phase Separation and Polymer Degradation. Vol. 120, pp. 207-266.

Kaetsu, I.: Radiation Synthesis of Polymeric Materials for Biomedical and Biochemical Applications. Vol. 105, pp. 81-98.
Kaji, K. see Kanaya, T.: Vol. 154, pp. 87-141.
Kajzar, F., Lee, K.-S., Jen, A.K.-Y.: Polymeric Materials and their Orientation Techniques for Second-Order Nonlinear Optics. Vol. 161, pp. 1-85.
Kakimoto, M. see Gaw, K. O.: Vol. 140, pp. 107-136.
Kaminski, W. and Arndt, M.: Metallocenes for Polymer Catalysis. Vol. 127, pp. 143-187.
Kammer, H. W., Kressler, H. and Kummerloewe, C.: Phase Behavior of Polymer Blends - Effects of Thermodynamics and Rheology. Vol. 106, pp. 31-86.
Kanaya, T. and Kaji, K.: Dynamcis in the Glassy State and Near the Glass Transition of Amorphous Polymers as Studied by Neutron Scattering. Vol. 154, pp. 87-141.
Kandyrin, L. B. and Kuleznev, V. N.: The Dependence of Viscosity on the Composition of Concentrated Dispersions and the Free Volume Concept of Disperse Systems. Vol. 103, pp. 103-148.
Kaneko, M. see Ramaraj, R.: Vol. 123, pp. 215-242.
Kang, E. T., Neoh, K. G. and Tan, K. L.: X-Ray Photoelectron Spectroscopic Studies of Electroactive Polymers. Vol. 106, pp. 135-190.

Karlsson, S. see Söderqvist Lindblad, M.: Vol. 157, pp. 139–161.

Kato, K. see Uyama, Y.: Vol. 137, pp. 1-40.

Kawaguchi, S. see Ito, K.: Vol. 142, p 129-178.

Kazanskii, K. S. and *Dubrovskii, S. A.*: Chemistry and Physics of „Agricultural" Hydrogels. Vol. 104, pp. 97-134.

Kennedy, J. P. see Jacob, S.: Vol. 146, pp. 1-38.

Kennedy, J. P. see Majoros, I.: Vol. 112, pp. 1-113.

Khokhlov, A., Starodybtzev, S. and *Vasilevskaya, V.*: Conformational Transitions of Polymer Gels: Theory and Experiment. Vol. 109, pp. 121-172.

Kiefer, J., Hedrick J. L. and *Hiborn, J. G.*: Macroporous Thermosets by Chemically Induced Phase Separation. Vol. 147, pp. 161-247.

Kilian, H. G. and *Pieper, T.*: Packing of Chain Segments. A Method for Describing X-Ray Patterns of Crystalline, Liquid Crystalline and Non-Crystalline Polymers. Vol. 108, pp. 49-90.

Kim, J. see Quirk, R.P.: Vol. 153, pp. 67-162.

Kim, K.-S. see Lin, T.-C.: Vol. 161, pp. 157-193.

Kippelen, B. and Peyghambarian, N.: Photorefractive Polymers and their Applications. Vol. 161, pp. 87-156.

Kishore, K. and *Ganesh, K.*: Polymers Containing Disulfide, Tetrasulfide, Diselenide and Ditelluride Linkages in the Main Chain. Vol. 121, pp. 81-122.

Kitamaru, R.: Phase Structure of Polyethylene and Other Crystalline Polymers by Solid-State ^{13}C/MNR. Vol. 137, pp 41-102.

Klee, D. and *Höcker, H.*: Polymers for Biomedical Applications: Improvement of the Interface Compatibility. Vol. 149, pp. 1-57.

Klier, J. see Scranton, A. B.: Vol. 122, pp. 1-54.

Klüppel, M. see Heinrich, G.: Vol. 160, pp 1-44.

Kobayashi, S., Shoda, S. and *Uyama, H.*: Enzymatic Polymerization and Oligomerization. Vol. 121, pp. 1-30.

Köhler, W. and *Schäfer, R.*: Polymer Analysis by Thermal-Diffusion Forced Rayleigh Scattering. Vol. 151, pp. 1-59.

Koenig, J. L. see Andreis, M.: Vol. 124, pp. 191-238.

Koike, T.: Viscoelastic Behavior of Epoxy Resins Before Crosslinking. Vol. 148, pp. 139-188.

Kokufuta, E.: Novel Applications for Stimulus-Sensitive Polymer Gels in the Preparation of Functional Immobilized Biocatalysts. Vol. 110, pp. 157-178.

Konno, M. see Saito, S.: Vol. 109, pp. 207-232.

Kopecek, J. see Putnam, D.: Vol. 122, pp. 55-124.

Koßmehl, G. see Schopf, G.: Vol. 129, pp. 1-145.

Kozlov, E. see Prokop, A.: Vol. 160, pp. 119-174.

Kramer, E. J. see Creton, C.: Vol. 156, pp. 53-135.

Kremer, K. see Baschnagel, J.: Vol. 152, pp. 41-156.

Kressler, J. see Kammer, H. W.: Vol. 106, pp. 31-86.

Kricheldorf, H. R.: Liquid-Cristalline Polyimides. Vol. 141, pp. 83-188.

Krishnamoorti, R. see Giannelis, E.P.: Vol. 138, pp. 107-148.

Kirchhoff, R. A. and *Bruza, K. J.*: Polymers from Benzocyclobutenes. Vol. 117, pp. 1-66.

Kuchanov, S. I.: Modern Aspects of Quantitative Theory of Free-Radical Copolymerization. Vol. 103, pp. 1-102.

Kuchanov, S. I.: Principles of Quantitive Description of Chemical Structure of Synthetic Polymers. Vol. 152, p. 157-202.

Kudaibergennow, S.E.: Recent Advances in Studying of Synthetic Polyampholytes in Solutions. Vol. 144, pp. 115-198.

Kuleznev, V. N. see Kandyrin, L. B.: Vol. 103, pp. 103-148.

Kulichkhin, S. G. see Malkin, A. Y.: Vol. 101, pp. 217-258.

Kulicke, W.-M. see Grigorescu, G.: Vol. 152, p. 1-40.

Kumar, M.N.V.R., Kumar, N., Domb, A.J. and *Arora, M.*: Pharmaceutical Polyme-ric Controlled Drug Delivery Systems. Vol. 160, pp. 45-118.

Kumar, N. see Kumar M.N.V.R.: Vol. 160, pp. 45-118.

Kummerloewe, C. see Kammer, H. W.: Vol. 106, pp. 31-86.
Kuznetsova, N. P. see Samsonov, G. V.: Vol. 104, pp. 1-50.Labadie, J. W. see Hergenrother, P. M.: Vol. 117, pp. 67-110.

Labadie, J. W. see Hedrick, J. L.: Vol. 141, pp. 1-44.
Labadie, J. W. see Hedrick, J. L.: Vol. 147, pp. 61-112.
Lamparski, H. G. see O´Brien, D. F.: Vol. 126, pp. 53-84.
Laschewsky, A.: Molecular Concepts, Self-Organisation and Properties of Polysoaps. Vol. 124, pp. 1-86.
Laso, M. see Leontidis, E.: Vol. 116, pp. 283-318.
Lazár, M. and RychlΩ, R.: Oxidation of Hydrocarbon Polymers. Vol. 102, pp. 189-222.
Lechowicz, J. see Galina, H.: Vol. 137, pp. 135-172.
Léger, L., Raphaël, E., Hervet, H.: Surface-Anchored Polymer Chains: Their Role in Adhesion and Friction. Vol. 138, pp. 185-226.
Lenz, R. W.: Biodegradable Polymers. Vol. 107, pp. 1-40.
Leontidis, E., de Pablo, J. J., Laso, M. and Suter, U. W.: A Critical Evaluation of Novel Algorithms for the Off-Lattice Monte Carlo Simulation of Condensed Polymer Phases. Vol. 116, pp. 283-318.
Lee, B. see Quirk, R.P: Vol. 153, pp. 67-162.
Lee, K.-S. see Kajzar, F.: Vol. 161, pp. 1-85.
Lee, Y. see Quirk, R.P: Vol. 153, pp. 67-162.
Leónard,, D. see Mathieu, H. J.: Vol. 162, pp. 1-35.
Lesec, J. see Viovy, J.-L.: Vol. 114, pp. 1-42.
Li, M. see Jiang, M.: Vol. 146, pp. 121-194.
Liang, G. L. see Sumpter, B. G.: Vol. 116, pp. 27-72.
Lienert, K.-W.: Poly(ester-imide)s for Industrial Use. Vol. 141, pp. 45-82.
Lin, J. and *Sherrington, D. C.*: Recent Developments in the Synthesis, Thermostability and Liquid Crystal Properties of Aromatic Polyamides. Vol. 111, pp. 177-220.
Lin, T.-C., Chung, S.-J., Kim, K.-S., Wang, X., He, G. S., Swiatkiewicz, J., Pudavar, H. E. and Prasad, P. N.: Organics and Polymers with High Two-Photon Activities and their Applications. Vol. 161, pp. 157-193.
Liu, Y. see Söderqvist Lindblad, M.: Vol. 157, pp. 139–161
López Cabarcos, E. see Baltá-Calleja, F. J.: Vol. 108, pp. 1-48.

Majoros, I., Nagy, A. and *Kennedy, J. P.*: Conventional and Living Carbocationic Polymerizations United. I. A Comprehensive Model and New Diagnostic Method to Probe the Mechanism of Homopolymerizations. Vol. 112, pp. 1-113.
Makhija, S. see Jaffe, M.: Vol. 117, pp. 297-328.
Malmström, E. see Hult, A.: Vol. 143, pp. 1-34.
Malkin, A. Y. and *Kulichkhin, S. G.*: Rheokinetics of Curing. Vol. 101, pp. 217-258.
Maniar, M. see Domb, A. J.: Vol. 107, pp. 93-142.
Manias, E., see Giannelis, E.P.: Vol. 138, pp. 107-148.
Mashima, K., Nakayama, Y. and *Nakamura, A.*: Recent Trends in Polymerization of a-Olefins Catalyzed by Organometallic Complexes of Early Transition Metals. Vol. 133, pp. 1-52.
Mathew, D. see Reghunadhan Nair, C.P.: Vol. 155, pp. 1-99.
Mathieu, H. J., Chevolot, Y, Ruiz-Taylor, L. and Leónard, D.: Engineering and Characterization of Polymer Surfaces for Biomedical Applications. Vol. 162, pp. 1-35.
Matsumoto, A.: Free-Radical Crosslinking Polymerization and Copolymerization of Multivinyl Compounds. Vol. 123, pp. 41-80.
Matsumoto, A. see Otsu, T.: Vol. 136, pp. 75-138.
Matsuoka, H. and *Ise, N.*: Small-Angle and Ultra-Small Angle Scattering Study of the Ordered Structure in Polyelectrolyte Solutions and Colloidal Dispersions. Vol. 114, pp. 187-232.
Matsushige, K., Hiramatsu, N. and *Okabe, H.*: Ultrasonic Spectroscopy for Polymeric Materials. Vol. 125, pp. 147-186.
Mattice, W. L. see Rehahn, M.: Vol. 131/132, pp. 1-475.
Mattice, W. L. see Baschnagel, J.: Vol. 152, p. 41-156.

Mays, W. see Xu, Z.: Vol. 120, pp. 1-50.

Mays, J.W. see Pitsikalis, M.: Vol.135, pp. 1-138.

McGrath, J. E. see Hedrick, J. L.: Vol. 141, pp. 1-44.

McGrath, J. E., Dunson, D. L., Hedrick, J. L.: Synthesis and Characterization of Segmented Poly-imide-Polyorganosiloxane Copolymers. Vol. 140, pp. 61-106.

McLeish, T.C. B., Milner, S. T.: Entangled Dynamics and Melt Flow of Branched Polymers. Vol. 143, pp. 195-256.

Mecerreyes, D., Dubois, P. and Jerôme, R.: Novel Macromolecular Architectures Based on Aliphatic Polyesters: Relevance of the „Coordination-Insertion" Ring-Opening Poly-merization. Vol. 147, pp. 1 -60.

Mecham, S. J. see McGrath, J. E.: Vol. 140, pp. 61-106.

Mikos, A. G. see Thomson, R. C.: Vol. 122, pp. 245-274.

Milner, S. T. see McLeish, T. C. B.: Vol. 143, pp. 195-256.

Mison, P. and Sillion, B.: Thermosetting Oligomers Containing Maleimides and Nadiimides End-Groups. Vol. 140, pp. 137-180.

Miyasaka, K.: PVA-Iodine Complexes: Formation, Structure and Properties. Vol. 108. pp. 91-130.

Miller, R. D. see Hedrick, J. L.: Vol. 141, pp. 1-44.

Monnerie, L. see Bahar, I.: Vol. 116, pp. 145-206.

Morishima, Y.: Photoinduced Electron Transfer in Amphiphilic Polyelectrolyte Systems. Vol. 104, pp. 51-96.

Morton M. see Quirk, R.P: Vol. 153, pp. 67-162.

Mours, M. see Winter, H. H.: Vol. 134, pp. 165-234.

Müllen, K. see Scherf, U.: Vol. 123, pp. 1-40.

Müller-Plathe, F. see Gusev, A. A.: Vol. 116, pp. 207-248.

Müller-Plathe, F. see Baschnagel, J.: Vol. 152, p. 41-156.

Mukerherjee, A. see Biswas, M.: Vol. 115, pp. 89-124.

Murat, M. see Baschnagel, J.: Vol. 152, p. 41-156.

Mylnikov, V.: Photoconducting Polymers. Vol. 115, pp. 1-88.

Nagy, A. see Majoros, I.: Vol. 112, pp. 1-11.

Nakamura, A. see Mashima, K.: Vol. 133, pp. 1-52.

Nakayama, Y. see Mashima, K.: Vol. 133, pp. 1-52.

Narasinham, B., Peppas, N. A.: The Physics of Polymer Dissolution: Modeling Approaches and Experimental Behavior. Vol. 128, pp. 157-208.

Nechaev, S. see Grosberg, A.: Vol. 106, pp. 1-30.

Neoh, K. G. see Kang, E. T.: Vol. 106, pp. 135-190.

Newman, S. M. see Anseth, K. S.: Vol. 122, pp. 177-218.

Nijenhuis, K. te: Thermoreversible Networks. Vol. 130, pp. 1-252.

Ninan, K.N. see Reghunadhan Nair, C. P.: Vol. 155, pp. 1-99.

Noid, D. W. see Otaigbe, J.U.: Vol. 154, pp. 1-86.

Noid, D. W. see Sumpter, B. G.: Vol. 116, pp. 27-72.

Novac, B. see Grubbs, R.: Vol. 102, pp. 47-72.

Novikov, V. V. see Privalko, V. P.: Vol. 119, pp. 31-78.

O'Brien, D. F., Armitage, B. A., Bennett, D. E. and *Lamparski, H. G.:* Polymerization and Domain Formation in Lipid Assemblies. Vol. 126, pp. 53-84.

Ogasawara, M.: Application of Pulse Radiolysis to the Study of Polymers and Polymerizations. Vol.105, pp. 37-80.

Okabe, H. see Matsushige, K.: Vol. 125, pp. 147-186.

Okada, M.: Ring-Opening Polymerization of Bicyclic and Spiro Compounds. Reactivities and Polymerization Mechanisms. Vol. 102, pp. 1-46.

Okano, T.: Molecular Design of Temperature-Responsive Polymers as Intelligent Materials. Vol. 110, pp. 179-198.

Okay, O. see Funke, W.: Vol. 136, pp. 137-232.

Onuki, A.: Theory of Phase Transition in Polymer Gels. Vol. 109, pp. 63-120.
Osad'ko, I.S.: Selective Spectroscopy of Chromophore Doped Polymers and Glasses. Vol. 114, pp. 123-186.
Otaigbe, J. U., Barnes, M. D., Fukui, K., Sumpter, B. G., Noid, D. W.: Generation, Characterization, and Modeling of Polymer Micro- and Nano-Particles. Vol. 154, pp. 1-86.
Otsu, T., Matsumoto, A.: Controlled Synthesis of Polymers Using the Iniferter Technique: Developments in Living Radical Polymerization. Vol. 136, pp. 75-138.

de Pablo, J. J. see Leontidis, E.: Vol. 116, pp. 283-318.
Padias, A. B. see Penelle, J.: Vol. 102, pp. 73-104.
Pascault, J.-P. see Williams, R. J. J.: Vol. 128, pp. 95-156.
Pasch, H.: Analysis of Complex Polymers by Interaction Chromatography. Vol. 128, pp. 1-46.
Pasch, H.: Hyphenated Techniques in Liquid Chromatography of Polymers. Vol. 150, pp. 1-66.
Paul, W. see Baschnagel, J.: Vol. 152, p. 41-156.
Penczek, P. see Batog, A. E.: Vol. 144, pp. 49-114.
Penelle, J., Hall, H. K., Padias, A. B. and *Tanaka, H.*: Captodative Olefins in Polymer Chemistry. Vol. 102, pp. 73-104.
Peppas, N. A. see Bell, C. L.: Vol. 122, pp. 125-176.
Peppas, N.A. see Hassan, C.M.: Vol. 153, pp. 37-65
Peppas, N. A. see Narasimhan, B.: Vol. 128, pp. 157-208.
Pet'ko, I. P. see Batog, A. E.: Vol. 144, pp. 49-114.
Pheyghambarian, N. see Kippelen, B.: Vol. 161, pp. 87-156.
Pichot, C. see Hunkeler, D.: Vol. 112, pp. 115-134.
Pieper, T. see Kilian, H. G.: Vol. 108, pp. 49-90.
Pispas, S. see Pitsikalis, M.: Vol. 135, pp. 1-138.
Pispas, S. see Hadjichristidis: Vol. 142, pp. 71-128.
Pitsikalis, M., Pispas, S., Mays, J. W., Hadjichristidis, N.: Nonlinear Block Copolymer Architectures. Vol. 135, pp. 1-138.
Pitsikalis, M. see Hadjichristidis: Vol. 142, pp. 71-128.
Pötschke, D. see Dingenouts, N.: Vol 144, pp. 1-48.
Pokrovskii, V. N.: The Mesoscopic Theory of the Slow Relaxation of Linear Macromolecules. Vol. 154, pp. 143-219.
Pospíšil, J.: Functionalized Oligomers and Polymers as Stabilizers for Conventional Polymers. Vol. 101, pp. 65-168.
Pospíšil, J.: Aromatic and Heterocyclic Amines in Polymer Stabilization. Vol. 124, pp. 87-190.
Powers, A. C. see Prokop, A.: Vol. 136, pp. 53-74.
Prasad, P. N. see Lin, T.-C.: Vol. 161, pp. 157-193.
Priddy, D. B.: Recent Advances in Styrene Polymerization. Vol. 111, pp. 67-114.
Priddy, D. B.: Thermal Discoloration Chemistry of Styrene-co-Acrylonitrile. Vol. 121, pp. 123-154.
Privalko, V. P. and *Novikov, V. V.*: Model Treatments of the Heat Conductivity of Heterogeneous Polymers. Vol. 119, pp 31-78.
Prokop, A., Hunkeler, D., Powers, A. C., Whitesell, R. R., Wang, T. G.: Water Soluble Polymers for Immunoisolation II: Evaluation of Multicomponent Microencapsulation Systems. Vol. 136, pp. 53-74.
Prokop, A., Hunkeler, D., DiMari, S., Haralson, M. A., Wang, T. G.: Water Soluble Polymers for Immunoisolation I: Complex Coacervation and Cytotoxicity. Vol. 136, pp. 1-52.
Prokop, A., Kozlov, E., Carlesso, G. and Davidsen, J.M.: Hydrogel-Based Colloidal Polymeric System for Protein and Drug Delivery: Physical and Chemical Characterization, Permeability Control and Applications. Vol. 160, pp. 119-174.
Pruitt, L. A.: The Effects of Radiation on the Structural and Mechanical Properties of Medical Polymers. Vol. 162, pp. 65-95.
Pudavar, H. E. see Lin, T.-C.: Vol. 161, pp. 157-193.

Pukánszky, B. and Fekete, E.: Adhesion and Surface Modification. Vol. 139, pp. 109-154.
Putnam, D. and Kopecek, J.: Polymer Conjugates with Anticancer Acitivity. Vol. 122, pp. 55- 124.

Quirk, R.P. and Yoo, T., Lee, Y., M., Kim, J. and Lee, B.: Applications of 1,1-Diphenylethylene Chemistry in Anionic Synthesis of Polymers with Controlled Structures. Vol. 153, pp. 67-162.

Ramaraj, R. and Kaneko, M.: Metal Complex in Polymer Membrane as a Model for Photosynthetic Oxygen Evolving Center. Vol. 123, pp. 215-242.
Rangarajan, B. see Scranton, A. B.: Vol. 122, pp. 1-54.
Ranucci, E. see Söderqvist Lindblad, M.: Vol. 157, pp. 139–161.
Raphaël, E. see Léger, L.: Vol. 138, pp. 185-226.
Reddinger, J. L. and Reynolds, J. R.: Molecular Engineering of π-Conjugated Polymers. Vol. 145, pp. 57-122.
Reghunadhan Nair, C.P., Mathew, D. and Ninan, K.N., : Cyanate Ester Resins, Recent Developments. Vol. 155, pp. 1-99.
Reichert, K. H. see Hunkeler, D.: Vol. 112, pp. 115-134.
Rehahn, M., Mattice, W. L., Suter, U. W.: Rotational Isomeric State Models in Macromolecular Systems. Vol. 131/132, pp. 1-475.
Reynolds, J.R. see Reddinger, J. L.: Vol. 145, pp. 57-122.
Richter, D. see Ewen, B.: Vol. 134, pp.1-130.
Risse, W. see Grubbs, R.: Vol. 102, pp. 47-72.
Rivas, B. L. and Geckeler, K. E.: Synthesis and Metal Complexation of Poly(ethyleneimine) and Derivatives. Vol. 102, pp. 171-188.
Robin, J. J. see Boutevin, B.: Vol. 102, pp. 105-132.
Roe, R.-J.: MD Simulation Study of Glass Transition and Short Time Dynamics in Polymer Liquids. Vol. 116, pp. 111-114.
Roovers, J., Comanita, B.: Dendrimers and Dendrimer-Polymer Hybrids. Vol. 142, pp 179-228.
Rothon, R. N.: Mineral Fillers in Thermoplastics: Filler Manufacture and Characterisation. Vol. 139, pp. 67-108.
Rozenberg, B. A. see Williams, R. J. J.: Vol. 128, pp. 95-156.
Ruckenstein, E.: Concentrated Emulsion Polymerization. Vol. 127, pp. 1-58.
Ruiz-Taylor, L. see Mathieu, H. J.: Vol. 162, pp. 1-35.
Rusanov, A. L.: Novel Bis (Naphtalic Anhydrides) and Their Polyheteroarylenes with Improved Processability. Vol. 111, pp. 115-176.
Russel, T. P. see Hedrick, J. L.: Vol. 141, pp. 1-44.
Rychlý, J. see Lazár, M.: Vol. 102, pp. 189-222.
Ryner, M. see Stridsberg, K. M.: Vol. 157, pp. 27–51.
Ryzhov, V. A. see Bershtein, V. A.: Vol. 114, pp. 43-122.

Sabsai, O. Y. see Barshtein, G. R.: Vol. 101, pp. 1-28.
Saburov, V. V. see Zubov, V. P.: Vol. 104, pp. 135-176.
Saito, S., Konno, M. and Inomata, H.: Volume Phase Transition of N-Alkylacrylamide Gels. Vol. 109, pp. 207-232.
Samsonov, G. V. and Kuznetsova, N. P.: Crosslinked Polyelectrolytes in Biology. Vol. 104, pp. 1-50.
Santa Cruz, C. see Baltá-Calleja, F. J.: Vol. 108, pp. 1-48.
Santos, S. see Baschnagel, J.: Vol. 152, p. 41-156.
Sato, T. and Teramoto, A.: Concentrated Solutions of Liquid-Christalline Polymers. Vol. 126, pp. 85-162.
Schäfer R. see Köhler, W.: Vol. 151, pp. 1-59.
Scherf, U. and Müllen, K.: The Synthesis of Ladder Polymers. Vol. 123, pp. 1-40.
Schmidt, M. see Förster, S.: Vol. 120, pp. 51-134.
Scholz, M.: Effects of Ion Radiation on Cells and Tissues. Vol. 162, pp. 97-158.
Schopf, G. and Koßmehl, G.: Polythiophenes - Electrically Conductive Polymers. Vol. 129, pp. 1-145.

Schweizer, K. S.: Prism Theory of the Structure, Thermodynamics, and Phase Transitions of Polymer Liquids and Alloys. Vol. 116, pp. 319-378.

Scranton, A. B., Rangarajan, B. and *Klier, J.*: Biomedical Applications of Polyelectrolytes. Vol. 122, pp. 1-54.

Sefton, M. V. and *Stevenson, W. T. K.*: Microencapsulation of Live Animal Cells Using Polycrylates. Vol.107, pp. 143-198.

Shamanin, V. V.: Bases of the Axiomatic Theory of Addition Polymerization. Vol. 112, pp. 135-180.

Sheiko, S. S.: Imaging of Polymers Using Scanning Force Microscopy: From Superstructures to Individual Molecules. Vol. 151, pp. 61-174.

Sherrington, D. C. see Cameron, N. R. , Vol. 126, pp. 163-214.

Sherrington, D. C. see Lin, J.: Vol. 111, pp. 177-220.

Sherrington, D. C. see Steinke, J.: Vol. 123, pp. 81-126.

Shibayama, M. see Tanaka, T.: Vol. 109, pp. 1-62.

Shiga, T.: Deformation and Viscoelastic Behavior of Polymer Gels in Electric Fields. Vol. 134, pp. 131-164.

Shim, H.-K., Jin, J.: Light-Emitting Characteristics of Conjugated Polymers. Vol. 158, pp. 191-241.

Shoda, S. see Kobayashi, S.: Vol. 121, pp. 1-30.

Siegel, R. A.: Hydrophobic Weak Polyelectrolyte Gels: Studies of Swelling Equilibria and Kinetics. Vol. 109, pp. 233-268.

Silvestre, F. see Calmon-Decriaud, A.: Vol. 207, pp. 207-226.

Sillion, B. see Mison, P.: Vol. 140, pp. 137-180.

Singh, R. P. see Sivaram, S.: Vol. 101, pp. 169-216.

Sinha Ray, S. see Biswas, M: Vol. 155, pp. 167-221.

Sivaram, S. and *Singh, R. P.*: Degradation and Stabilization of Ethylene-Propylene Copolymers and Their Blends: A Critical Review. Vol. 101, pp. 169-216.

Söderqvist Lindblad, M., Liu, Y., Albertsson, A.-C., Ranucci, E., Karlsson, S.: Polymer from Renewable Resources. Vol. 157, pp. 139–161

Starodybtzev, S. see Khokhlov, A.: Vol. 109, pp. 121-172.

Stegeman, G. I.: see Canva, M.: Vol. 158, pp. 87-121.

Steinke, J., Sherrington, D. C. and *Dunkin, I. R.*: Imprinting of Synthetic Polymers Using Molecular Templates. Vol. 123, pp. 81-126.

Stenzenberger, H. D.: Addition Polyimides. Vol. 117, pp. 165-220.

Stevenson, W. T. K. see Sefton, M. V.: Vol. 107, pp. 143-198.

Stridsberg, K. M., Ryner, M., Albertsson, A.-C.: Controlled Ring-Opening Polymerization: Polymers with Designed Macromoleculars Architecture. Vol. 157, pp. 27-51.

Suematsu, K.: Recent Progress of Gel Theory: Ring, Excluded Volume, and Dimension. Vol. 156, pp. 136-214.

Sumpter, B. G., Noid, D. W., Liang, G. L. and *Wunderlich, B.*: Atomistic Dynamics of Macromolecular Crystals. Vol. 116, pp. 27-72.

Sumpter, B. G. see Otaigbe, J.U.: Vol. 154, pp. 1-86.

Sugimoto, H. and *Inoue, S.*: Polymerization by Metalloporphyrin and Related Complexes. Vol. 146, pp. 39-120.

Suter, U. W. see Gusev, A. A.: Vol. 116, pp. 207-248.

Suter, U. W. see Leontidis, E.: Vol. 116, pp. 283-318.

Suter, U. W. see Rehahn, M.: Vol. 131/132, pp. 1-475.

Suter, U. W. see Baschnagel, J.: Vol. 152, p. 41-156.

Suzuki, A.: Phase Transition in Gels of Sub-Millimeter Size Induced by Interaction with Stimuli. Vol. 110, pp. 199-240.

Suzuki, A. and *Hirasa, O.*: An Approach to Artifical Muscle by Polymer Gels due to Micro-Phase Separation. Vol. 110, pp. 241-262.

Swiatkiewicz, J. see Lin, T.-C.: Vol. 161, pp. 157-193.

Tagawa, S.: Radiation Effects on Ion Beams on Polymers. Vol. 105, pp. 99-116.

Tan, K. L. see Kang, E. T.: Vol. 106, pp. 135-190.

Tanaka, H. and *Shibayama, M.*: Phase Transition and Related Phenomena of Polymer Gels. Vol. 109, pp. 1-62.

Tanaka, T. see Penelle, J.: Vol. 102, pp. 73-104.

Tauer, K. see Guyot, A.: Vol. 111, pp. 43-66.

Teramoto, A. see Sato, T.: Vol. 126, pp. 85-162.

Terent´eva, J. P. and *Fridman, M. L.*: Compositions Based on Aminoresins. Vol. 101, pp. 29-64.

Theodorou, D. N. see Dodd, L. R.: Vol. 116, pp. 249-282.

Thomson, R. C., Wake, M. C., Yaszemski, M. J. and *Mikos, A. G.*: Biodegradable Polymer Scaffolds to Regenerate Organs. Vol. 122, pp. 245-274.

Tokita, M.: Friction Between Polymer Networks of Gels and Solvent. Vol. 110, pp. 27-48.

Tries, V. see Baschnagel, J:. Vol. 152, p. 41-156.

Tsuruta, T.: Contemporary Topics in Polymeric Materials for Biomedical Applications. Vol. 126, pp. 1-52.

Uyama, H. see Kobayashi, S.: Vol. 121, pp. 1-30.

Uyama, Y: Surface Modification of Polymers by Grafting. Vol. 137, pp. 1-40.

Varma, I. K. see Albertsson, A.-C.: Vol. 157, pp. 99-138.

Vasilevskaya, V. see Khokhlov, A.: Vol. 109, pp. 121-172.

Vaskova, V. see Hunkeler, D.: Vol.:112, pp. 115-134.

Verdugo, P.: Polymer Gel Phase Transition in Condensation-Decondensation of Secretory Products. Vol. 110, pp. 145-156.

Vettegren, V. I.: see Bronnikov, S. V.: Vol. 125, pp. 103-146.

Viovy, J.-L. and *Lesec, J.*: Separation of Macromolecules in Gels: Permeation Chromatography and Electrophoresis. Vol. 114, pp. 1-42.

Vlahos, C. see Hadjichristidis, N.: Vol. 142, pp. 71-128.

Volksen, W.: Condensation Polyimides: Synthesis, Solution Behavior, and Imidization Characteristics. Vol. 117, pp. 111-164.

Volksen, W. see Hedrick, J. L.: Vol. 141, pp. 1-44.

Volksen, W. see Hedrick, J. L.: Vol. 147, pp. 61-112.

Wake, M. C. see Thomson, R. C.: Vol. 122, pp. 245-274.

Wandrey C., Hernández-Barajas, J. and *Hunkeler, D.*: Diallyldimethylammonium Chloride and its Polymers. Vol. 145, pp. 123-182.

Wang, K. L. see Cussler, E. L.: Vol. 110, pp. 67-80.

Wang, S.-Q.: Molecular Transitions and Dynamics at Polymer/Wall Interfaces: Origins of Flow Instabilities and Wall Slip. Vol. 138, pp. 227-276.

Wang, T. G. see Prokop, A.: Vol. 136, pp.1-52; 53-74.

Wang, X. see Lin, T.-C.: Vol. 161, pp. 157-193.

Whitesell, R. R. see Prokop, A.: Vol. 136, pp. 53-74.

Williams, R. J. J., Rozenberg, B. A., Pascault, J.-P.: Reaction Induced Phase Separation in Modified Thermosetting Polymers. Vol. 128, pp. 95-156.

Winter, H. H., Mours, M.: Rheology of Polymers Near Liquid-Solid Transitions. Vol. 134, pp. 165-234.

Wu, C.: Laser Light Scattering Characterization of Special Intractable Macromolecules in Solution. Vol 137, pp. 103-134.

Wunderlich, B. see Sumpter, B. G.: Vol. 116, pp. 27-72.

Xiang, M. see Jiang, M.: Vol. 146, pp. 121-194.

Xie, T. Y. see Hunkeler, D.: Vol. 112, pp. 115-134.

Xu, Z., Hadjichristidis, N., Fetters, L. J. and *Mays, J. W.*: Structure/Chain-Flexibility Relationships of Polymers. Vol. 120, pp. 1-50.

Yagci, Y. and *Endo, T.*: N-Benzyl and N-Alkoxy Pyridium Salts as Thermal and Photochemical Initiators for Cationic Polymerization. Vol. 127, pp. 59-86.

Yannas, I. V.: Tissue Regeneration Templates Based on Collagen-Glycosaminoglycan Copolymers. Vol. 122, pp. 219-244.

Yang, J. S. see Jo, W. H.: Vol. 156, pp. 1-52.

Yamaoka, H.: Polymer Materials for Fusion Reactors. Vol. 105, pp. 117-144.

Yasuda, H. and *Ihara, E.*: Rare Earth Metal-Initiated Living Polymerizations of Polar and Nonpolar Monomers. Vol. 133, pp. 53-102.

Yaszemski, M. J. see Thomson, R. C.: Vol. 122, pp. 245-274.

Yoo, T. see Quirk, R.P.: Vol. 153, pp. 67-162.

Yoon, D. Y. see Hedrick, J. L.: Vol. 141, pp. 1-44.

Yoshida, H. and *Ichikawa, T.*: Electron Spin Studies of Free Radicals in Irradiated Polymers. Vol. 105, pp. 3-36.

Zhou, H. see Jiang, M.: Vol. 146, pp. 121-194.

Zubov, V. P., Ivanov, A. E. and *Saburov, V. V.*: Polymer-Coated Adsorbents for the Separation of Biopolymers and Particles. Vol. 104, pp. 135-176.

Subject Index

α-particles   123
Accelerated aging   83
Acrylic acid   41, 42, 43, 44, 50
Amorphous track structure   142
Angle resolved XPS (ARXPS)
Antimicrobial   53
APC   15
AR-XPS   23

Basepairs   103
Biodegradable polymers   81
Biomaterials   3
Biomimicking Polymers   14
Biophysical models   140
Bone cement   91
Braggpeak   118
Build-up effect   118
Bystander effect   101, 136

Cell death   107
Cell division   100
Cell membrane   99
Cell nucleus   99
Chain scission   73
CHO   124
Chromatid breaks   134
Chromosome aberrations   105
Chromosome territories   145
Chromosomes   100
Clusters of damage   133
Clusters of DSB   133
Collagen   58
Complex damage   103, 123
Contact angle   50
Contact Angle Measurements   11
Correlated damage   121
Crosslinking   73, 87
Cytoplasm   99

Deletions   111
Depth dose distribution   117
Dicentric chromosomes   105
Differentiation   101
Dilution assay   107
Direct effect   102
Disintegration   134
Division cycle   100
DNA   99
DNA content   101
DNA damage processing   136
DNA fragments   104
DNA replication   100
DNA-protein crosslinks   103
Double strand breaks (DSB)   102
Drug   55
DSB   103

Early responding tissues   112
ECM   57
Exchange type aberrations   134

Fatigue crack propagation   85
Fidelity of the rejoining   133
Fluorescence in situ hybridization
   (FISH)   106
Fluropolymers   78
Fracture toughness   92
Free radical reactions   72
Functional subunits   112

Gel electrophoresis   103
Glycoengineering   24
Graft modification   39

HEMA   42, 47, 48, 49, 53
Hydrogels   81
Hypoxic cell fractions   113

Implant   55
Implantable devices   66
Inactivation cross section   129
Indirect effect   102
Infection   67
Information depth of XPS   7
Infrared spectra   84
Ionizing radiation   65, 68, 82

Lactose aryl diazirine   26
Late responding tissues   112
Latency period   112
Lesion interaction   140
Linear energy transfer (LET)   114
Linear-quadratic approach   107
Local effect model (LEM)   142

MAD-Gal   25, 26, 29
Mechanical properties   65
Medical polymers   65, 77
Methaacrylic acid   53
Methyl Acrylate (MA)   15
Methyl methacrylate (MMA)   15
Microbeam facilities   136
Microdosimetry   140
Misrejoining   105
Misrepair   105
Mitosis   100
Mitotic death   107
Molecular weight changes   69
–, environmental effects   72
Mutations   111

Neutral Red (NR)   30, 31
Neutrons   119
Nuclear interaction   118
Nylons   80

Orthopedics   82
Overkill effects   128
Oxidation embrittlement   84
Oxygen enhancement ratio (OER)   111

P(MMA:MA:APC)   14, 17, 22
P21-protein   136
Particle track   113
Particle traversals   127
PC copolymer   14
PCPUR   14, 17, 21
PEO   46, 55
PET   41, 53, 58
Phosphorylcholine   14
Plasma   40
Point mutations   111

Polyacrylates   78
Polyesters   81
Polypropylene   78
Polystyrene   24
Polyurethanes   80
Positron emission tomography (PET)   147
Premature chromosome condensation
   (PCC)   107
Protons   123
PS   27
PTFE   13

Radial dose profile   114
Radiation   45
Radiation   65
Radiation dose, effect of   76
Radioprotectors   111
Radiosensitivity   101
Radiosensitizers   111
Raster scan system   147
Rejoining   105
Rejoining kinetics   133
Relative Biological Effectiveness
   (RBE)   120
Reoxygenation   140
Repair   103, 108
Repair capacity   121
Residual damage   105, 133

Secondary electrons   113
Signaling   101
Single Strand Break   102
Smooth muscle cells   59
SSB   103
Static SIMS   9
Stem cells   100, 112
Sterilization   66
Stopping power   114
Strand break induction   103
Sublethal damage   108
Surface   50
Surface Charakterization   3
Survival probabilities   107
Suture   53

Theory of dual radiation action
   (TDRA)   140
Therapeutic gain   148
Time-of-Flight Secondary Ion Mass Spec-
   trometry (ToF-SIMS)   8
Tissue Engineering   57
ToF-SIMS   23, 26
Track radius   116
Trajectory   113

Treatment planning   143
Tumor cure   139
Tumor growth delay   139
Tumor therapy   139

UHMWPE   75
Ultra violet   44
Urothelial cells   59

V79   124

Wear resistance   84, 89

XPS   17, 26
XPS spectrometer   5
X-ray Photoelectron Spectrometry
   (XPS)   4
XRS   124

Printing (Computer to Plate): Saladruck Berlin
Binding: Stürtz AG, Würzburg